U0177800

[英国]布赖恩·查尔斯沃思 德博拉·查尔斯沃思 著 舒中亚 译

# 进化

牛津通识读本·

Evolution

A Very Short Introduction

译林出版社

图书在版编目(CIP)数据
进化/(英)查尔斯沃思(Charlesworth, B.),(英)查尔斯沃思(Charlesworth, D.)著,舒中亚译.—南京:译林出版社,2015.3(2021.1重印)
(牛津通识读本)
书名原文:Evolution: A Very Short Introduction
ISBN 978-7-5447-5110-0

I.①进… II.①查… ②查… ③舒… III.①生物 - 进化 IV.①Q11

中国版本图书馆CIP数据核字(2014)第265006号



著作权合同登记号 图字: 10-2014-197 号

**进化 [英国] 布赖恩·查尔斯沃思 德博拉·查尔斯沃思 / 著 舒中亚 / 译**

责任编辑 许 丹 何本国
责任印制 董 虎

原文出版 Oxford University Press, 2003
出版发行 译林出版社
地 址 南京市湖南路1号A楼
邮 箱 yilin@yilin.com
网 址 www.yilin.com
市场热线 025-86633278
排 版 南京展望文化发展有限公司
印 刷 江苏凤凰通达印刷有限公司
开 本 635毫米 × 889毫米 1/16
印 张 18.25
插 页 4
版 次 2015年3月第1版
印 次 2021年1月第7次印刷
书 号 ISBN 978-7-5447-5110-0
定 价 39.00元

# 序言

周忠和

译林出版社翻译了牛津大学出版社2003年出版的《进化》（*Evolution: A Very Short Introduction*）一书，我由衷感到高兴。能够被邀请作序，虽诚惶诚恐，也深感荣幸，权且当作是一次学习的经历，因此通读了译稿，写下几句读书心得，与读者分享。

这是一本扼要介绍有关生物进化知识的著作，权威性毋庸置疑，而且十分通俗、有趣。它没有试图包括有关进化的各个重要方面的最新研究进展，但还是涉及了有关进化的一些最基本的问题。本书对与进化有关的许多生物学的基本概念，如物种、基因、突变等都做了很好的诠释，同时也概要总结了支持进化论的方方面面的证据，包括来自地质学和古生物学的证据。

本书讨论了许多与我们日常生活相关的或者一些容易令人产生兴趣的问题，而且列举了大量的例子。譬如，基因突变如何导致一些疾病？抗生素为什么不能过量使用？怎样从进化的角度来理解不同物种衰老的过程？社会性的动物中，为什么会进化出失去了生殖能力的个体？

本书一个鲜明的特点是应用了大量的比喻，来解释一些普通读者较为难懂的生物学概念。例如，通过抛硬币解释遗传漂变；用英语等有关联的语言拼写的不同，来形容不同物种同一基因序列的异同；认为不同语言差异的程度与它们分离的时间有一定的关联，同样的道理也适用于DNA（脱氧核糖核酸）序列的比较。作者还把简单原始生命今天依然成功延续比喻为老的但依然实用的工具，譬如现代办公室内电脑旁边的锤子。

本书专门有一章解释“适应与自然选择”。考虑到公众通常很容易对“适应”一词产生的误解和争议，作者用了比较多的篇幅来解释这一概念的内涵，解释充分而得体，并且告诉读者进化生物学意义上的“适应”（譬如，性选择产生的孔雀雍容华丽的尾巴）与我们日常生活中经常理解的“适应”是有差异的。此外，生物的进化并没有前瞻性或者预见性，冷酷无情的自然选择只是考虑眼前，而我们日常生活中理解的“适应”可能还包括了未来的考虑。

本书的最后一章“一些难题”也很有用意。对普通读者来说，进化论总是存在这样或那样不太容易理解的方面。生物复杂结构（如脊椎动物的眼睛）的进化恐怕就是其中的“难题”之一。对复杂结构的进化的理解之所以被称为“难题”，并不是因为理论上遇到了什么样的挑战，而是我们缺少足够多的过渡性的中间阶段的证据。至于其他的难题，譬如人类意识的起源和进化的问题，恐怕不仅难在证据的获取，而且如同作者所说的那样，“我们甚至很难清晰地表述这个问题的本质，因为众所周知意识是很难准确定义的”。

在书的结尾，与其说是后记，还不如说是一些富有哲理的思考。例如，“人类的多数变异都存在于同一区域群体的个体间，不同群体间的差异则少得多。因此，种族是同质的、彼此独立的存在这种想法是毫无道理的，而某个种族具有遗传上的‘优越性’这一说法更是无稽之谈”。 在进化过程中有进步吗？本书作者提供的答案是有保留的肯定。显然，复杂的生物来自不太复杂的生物，从原核单细胞到鸟类和哺乳动物的演化过程似乎表明了这一现象；然而，自然选择并没有暗示这一过程是不可避免的，细菌显然还是最丰富和成功的生命形式之一，复杂性减退的例子比比皆是。由此，我想到了“evolution”一词的中文翻译问题。究竟翻译为“进化”还是“演化”更好呢？尽管我个人更偏爱“演化”，相信读者看完本书之后也会有自己的偏爱和选择。

献给约翰·梅纳德·史密斯

# 致谢

感谢牛津大学出版社的谢利·考克斯与埃玛·西蒙斯，她们建议我们写作本书，并担任编辑。同样感谢海伦·博斯威克、简·查尔斯沃思与约翰·梅纳德·史密斯，他们对本书的初稿进行了阅读与点评。当然，本书存在任何错误，由我们负责。

# 目录

第一章
# 引言

匍匐者兮，
与我何异；
彼猿吾人，
兄弟血亲。

——托马斯·哈代《饮酒歌》

科学界存在着这样一个共识：地球是一颗行星，它环绕着一颗非常典型的恒星运动，这颗恒星是银河系中数以亿计的恒星之一，银河系又是不断扩张的浩瀚宇宙中数以亿计的星系之一，而宇宙起源于约140亿年前。大约46亿年前，由于尘埃与气体的引力凝聚过程，地球产生了；这个过程也产生了太阳以及其他太阳系的行星。早于35亿年前，在纯粹的化学变化中产生了能自我复制的分子们，所有现存生命都是它们的后代。在达尔文所谓的"后代渐变"过程中，生命渐次形成，通过一个枝蔓丛生的谱系——生命之树——彼此关联。我们人类与黑猩猩及大猩猩的亲缘关系最为紧密，在600万—700万年前，我们与它们有着共同的祖先。我们所属的哺乳类动物，与现存的爬行类动物在约3亿年前也具有共同的祖先。所有的脊椎动物（哺乳类、鸟类、爬行类、两栖类、鱼类）都能追根溯源到一种生活在5亿多年前、小小的像鱼一样、缺少脊柱的生物。再往前追溯，辨明动物、植物以及微生物的类群之间的关系变得愈加困难。但是，就像我们将要看到的那样，在它们的遗传物质中，有清晰而鲜明的印迹表明它们有共同的祖先。

距今不到450年，所有的欧洲学者还都相信地球是宇宙的中心，而宇宙的大小最多不过几百万英里，太阳以及其他星球都围绕着地球这个中心运动。距今不到250年，他们还认为宇宙是在6000年前被创造出来的，而它自创世之后就再无本质变化，尽管当时人们已经了解到地球与其他行星一样，是围绕着太阳转动，也广泛接受了宇宙比他们之前所了解的要大得多这一事实。距今不到150年，科学家们普遍接受了地球当前的状态是由至少数千万年的地质变迁形成的这一观点，但是生命是上帝的特别创造这一观念依然是主流思想。

在不到500年间，科学方法大行其道，我们通过实验与观测进行推理，而不再求助于宗教权威或者统治权威，这彻底改变了我们对于人类起源以及人与宇宙关系的观念。在开启了一个具有内在魅力的全新世界的同时，科学也深刻影响了哲学与宗教。科学的发现表明了人类是客观力量的产物，而我们所居住的世界只是浩瀚悠久的宇宙一个很小的组成部分。基于上述观点，我们才能探索了解宇宙——这是整个科学研究计划的基础假设，无论科学家们自身有着怎样的宗教或哲学信仰。

科学研究计划取得了令人瞩目的成功，这是毋庸置疑的，特别是在20世纪这个恐怖事件在人类社会中频频发生的时代。科学的影响或许间接地推动了这些事件——一方面通过大规模工业社会的兴起所触发的社会变革，另一方面通过对传统信仰体系的削弱侵蚀。然而，我们也可以说，人们本可以运用理性阻止人类历史上诸多悲剧的发生，20世纪的灾难源自理性的缺失而非理性的失败。正确运用科学来认识我们所生存的世界，这是人类未来的唯一希望。

进化论的研究已经揭示了我们与地球上栖息的其他物种之间紧密的联系，只有对这种联系保持尊重，才能避免全球性的大灾难的发生。自140多年前达尔文与华莱士发表本领域的首批著作以来，生物进化论蓬勃发展。本书的目的在于向大众读者们介绍一些进化生物学中最重要也最基础的发现、概念以及规程。进化论为整个生物学提供了一套统一的法则，也阐明了人类与宇宙及人与人之间的联系。此外，进化论的许多方面还有实际的价值，比如当前紧迫的医学问题，正是由于细菌对抗生素、艾滋病病毒对抗病毒药物快速进化出的抗药性所导致的。

在本书中，我们首先介绍进化论的主要因果过程（第二章）；第三章介绍了一些基本的生物背景知识，同时也展示了我们如何在进化层面理解生物之间的相似性；第四章描述了那些源自地球历史、源自现存物种地理分布样态的进化论证据；第五章关注自然选择之下的适应性进化；第六章则关注新物种的进化以及物种间差异的进化；在第七章中，我们将讨论一些对进化理论来说看似困难的问题；第八章则是一个简要的总结。

第二章
# 进化的过程

为了对地球上的生命有所了解，我们需要知道动物（包括人类）、植物以及微生物是如何运作的，并最终归结到构成它们运作基础的分子过程层面。这是生物学的"怎么样（how）"问题；在过去的一个世纪里，对于这个问题已有大量的研究，并取得了令人瞩目的进步。研究成果表明，即便是能够独立生存的最简单生物——细菌细胞，也是一台无比精密复杂的机器，拥有成千上万种不同的蛋白质分子；它们协同作用，提供细胞生存所必须的功能，并分裂产生两个子细胞（见第三章）。在更高等的生物如苍蝇或人类中，这种复杂性还会增大。上述生物的生命从一个由精子与卵细胞融合形成的单细胞开始，然后发生一系列受到精密调控的细胞分裂过程，与之相伴的是分裂产生的细胞分化成为多种不同的形态。发育的过程最终产生的是由不同组织与器官构成、具有高度有序结构、能够完成精细行为的成熟生物。我们对形成这种结构与功能复杂性的分子机理的了解正在快速进步。尽管还有许多尚待解决的问题，生物学家们相信，即使是生物中最为复杂的特性，比如人类的意识，也是化学与物理过程运行的反映，而这些过程能够被科学方法分析及探索。

从单个蛋白分子的结构与功能，到人类大脑的组成，在各级结构中我们都能看到许多**适应**的例证，这种结构对功能的适应与人类设计的机器有着异曲同工之妙（见第五章）。我们同样能看到，不同的物种具有相互迥异的特征，这些特征通常清晰地反映了它们对于栖息环境的适应。这些观察的结果引出了生物学的“为什么（why）”问题，涉及那些让生物体成为它们如今状态的过程。在“进化”的概念出现之前，大部分的生物学家在回答这一问题时，可能都会归因于造物主。“适应”这个词是18世纪英国的神学家引入的，他们认为，生物体特征中的精心设计的表象证明了一个超自然的设计师的存在。尽管这个理论被18世纪中叶的哲学家大卫·休谟证明具有逻辑缺陷，但在其他可供选择的可靠理论出现前，它依然在人们的思想中占有一席之地。

进化论思想引出了一系列自然过程，它们能够解释生物物种庞大的多样性，以及那些使生物较好适应栖息环境的特征，而不用诉诸超自然力量。这些解释自然也适用于人类本身的起源，这使得生物进化论成为一门最引人争议的科学。然而，如果我们不带任何偏见来看待这些问题，可以认为，支持进化是一个真实存在的过程的证据与其他确立多年的科学理论，如物质的分子特性（见第三、四章）一样，非常坚实可靠。有关进化的成因，我们同样有一系列已被充分验证的理论。不过，与所有健康发展的科学一样，在进化论中，同样存在尚待解决的问题，同时随着了解的深入，许多新的问题也在不断涌现（见第七章）。

生物的进化包括随着时间的推移生物种群特征所发生的变化。这种变化的时间尺度与大小的波动范围非常大。进化的研究可以完成于一个人的一生中，在此期间单一的特征发生了简单改变，例如为了控制细菌感染而广泛使用青霉素，在几年之内对青霉素有拮抗作用的菌株出现频率将会升高（第五章会讨论这一问题）。在另一个极端，进化也包括重要的新物种诞生这样的事件，这也许将花费几百万年的时间，需要许多不同特征的改变，例如从爬行动物向哺乳动物的转变（见第四章）。查尔斯·达尔文与阿尔弗雷德·拉塞尔·华莱士这两位进化论学说创始人的一个关键见解就是：各个层次的变化都可能包含同样类型的过程。进化方面的重要变化主要反映的是相同类型的微小改变经过长时间的积累造成的变化。

进化方面的改变最终要依靠生物体出现新的变化形态：**突变**。突变是由遗传物质的稳定变化造成的，由亲本传递给子代。实验遗传学家们已经研究了许多不同生物中几乎所有能够想到的特征的突变，医学遗传学家们业已列明人类种群中出现的数以千计的突变类型。生物表观特征上的突变结果差异很大。有一些突变并没有可观察到的表型，只是由于现在已经可以直接对遗传物质结构进行研究，人们才觉察到它们的存在（我们在第三章中将描述这一点）；另一些突变则是在某个简单特性上具有相对较小的影响，例如眼睛的颜色由棕色变成蓝色，某些细菌获得了针对某种抗生素的抗药性，或是果蝇体侧刚毛数量发生了改变。某些突变则对于生物发育具有极其显著的影响，例如黑腹果蝇的一种突变使得它的头部本该长触角的地方长出了一条腿。任何特殊类型的新突变的出现都是一个小概率事件，大概频率为在一代里10万个之中才出现一个，甚至还要更少。突变造成了一个特征的改变，例如抗生素耐药性，最初发生在单个个体之中，通常在许多代里这些变化被限制在一个很小的比例。为了达到进化方面的改变，需要有其他过程引发它在种群中频率的上升。

**自然选择**是进化改变的过程中最重要的一步，这些改变包括生物的结构、功能以及行为等方面（见第五章）。在1858年发表于《林奈学会议程学报》的论文中，达尔文与华莱士通过以下观点详述了他们的自然选择进化理论：

- 一个物种会产生大量后代，远超出能够正常存活到成熟期及繁殖期的数量，因此存在着**生存竞争**。
- 在种群的诸多特征中存在着个体变异，其中的一些可能会影响个体生存与繁殖的能力。因此某一代中成功繁殖的亲本可能与种群整体存在不同。
- 这些变异中的很大一部分可能具有**遗传组分**，因此成功亲本的子代特征将与上代的种群不同，而更接近于它们的亲本。

如果这个过程在每一代间继续，种群将出现渐进式的转变，由此与更强生存能力或更高繁殖成功率相关的特征的出现频率将随时间变化而升高。这种特征的改变起源于突变，但是影响单一特征的突变在任何时间都会出现，无论它是否会被自然选择所青睐。事实上，大部分的突变或是对于生物体没有影响，或是将降低生物生存或者繁殖的能力。

对于这种提高了生存或繁殖成功率的变异体而言，它的频率上升过程解释了适应性特征的进化，因为更强壮的身体或更好的表现通常能够提高个体生存或繁殖的成功率。当一个种群处于多变的环境中，这种变化过程尤其可能发生；在这种环境下，相较于那些已经被自然选择所确定的特性而言，一系列多少有些不一样的特征更容易受到青睐。正如达尔文在1858年所写的那样：

> 但是一旦外部环境改变……现在，每个个体都必须在竞争中寻求生存，任何一个能够使得个体更好适应新环境的结构、习性或本能上的微小变异，都将对个体的活力与健康造成影响，这是毋庸置疑的。在竞争中，这样的个体有着更好的生存**机会**；而遗传了这些变异——尽管如此微小——的后代，也同样具有更好的生存**机会**。年复一年，出生个体多于存活个体；天平中最细小的颗粒最终会决定谁将死亡，而谁又将生存。一手是自然选择，一手是死亡，在一千代之后，没有人能对它造成的影响视若无睹……

然而，同时存在另一种重要的进化改变机制，它解释了物种如何同样能够在对个体的生存或者繁殖成功率几无影响的性状上产生不同，这种机制因此不服从自然选择理论。正如我们将在第六章看到的那样，这种机制在遗传物质大类上的改变中尤其可能存在，这些改变对于机体的结构或功能几无影响。即使存在**选择中性**变异，因此通常情况下不同个体的生存或繁殖不存在差异，子代也依然可能与亲代存在细微差别。这是因为，在缺少自然选择的情况下，子代种群的基因是从亲代种群基因抽取的一个随机样本。真实种群的大小是有限的，于是子代种群的构成将与亲代存在随机差异，正如我们在抛10次硬币时，不会期望正好获得5次正面和5次背面。

这种随机变化的过程叫作**遗传漂变**。即使是最大的生物种群，例如细菌的种群，也是有限的，因此遗传漂变总是能够起作用。

突变、自然选择与遗传漂变的随机过程共同导致了种群组成的改变。在经历了足够长的一段时间后，这种累积效应改变了种群的基因组成，于是使得物种的特征与其祖先有了极大的差别。

我们在前文中提到了生命的多样性，这种多样性反映在了现存数量庞大的物种上。（有更多的物种在过去的年代里曾经存在过，但是正如第四章将描述的，灭绝是几乎所有物种的宿命。）新物种如何进化无疑是一个重要的问题，我们将在第六章中进行讨论。要定义“物种”这个词非常困难，想要在同一物种的种群与不同物种的种群间划出清晰的界线，有时也很困难。从进化角度看，当进行有性生殖的两个种群的生物体之间无法杂交，由此它们的进化轨迹完全独立，则可以认为它们是不同的物种，这种说法是有道理的。因此，居住在世界上不同地方的人类种群毫无疑问属于同一物种，因为如果有其他地区的移民到来，他们之间不存在杂交繁殖障碍。这种移民行为有助于防止同一物种的不同种群间的基因组成差异过大。与之相反，黑猩猩与人类显然就是不同物种，因为居住在同一地区的人类与黑猩猩之间不能够进行杂交繁殖。正如我们将在后文中提到的，人类与黑猩猩在遗传物质的组成上的差异同样要比人类本身之间的差异大得多。一个新物种的形成必然包括关联种群间杂交繁殖障碍的进化。一旦这种障碍形成，种群的发展将在突变、选择及遗传漂变的影响下产生分化。这种分化的过程最终导致生物的多样性。如果我们理解混合生殖障碍如何进化，种群又如何在之后发生分化，我们将理解物种的起源。

在进化论的这些观点的支持下，数量庞大的生物数据变得逐渐明朗。同时，正如天文学家与物理学家模拟恒星、行星、分子及原子的行为以求更彻底地了解它们，并给予自己的理论精细的检验，能够进行精确模拟的数学理论的发展也使得进化论有了坚实的基础。在更加具体地描述进化论的机制（但省略数学过程）之前，我们将在下面两章中展示进化论如何使得众多不同类型的生物发现变得有意义，与神创论的难以自圆其说形成鲜明对比。

第三章

# 进化的证据：生物间的相似与差异

进化论对生命的多样性做出了解释，其中包括动物、植物、微生物的不同物种间众所周知的差异；同时也解释了它们最基础的相似性。这些相似性通常在外部可见的特征这一表面层级上较为明显，同时也延伸至显微结构与生化功能中最精密的细部。我们将在本书的后文（第六章）中对生物的多样性进行讨论，同时阐述进化论如何解释“青出于蓝而胜于蓝”这一现象。但是，本章我们将着眼于生物的整体。此外，我们将介绍许多基本的生物学常识，后文的几章内容，正是建立在这些基本常识之上。

## 不同物种类别间的相似性

生物——即使是截然不同的生物——之间，在各种层面上都存在相似性。从我们熟悉的、外形上可见的相似，到更为深远的生命周期的相似，以及遗传物质结构的相似。即使在两种有着天壤之别的物种，如我们人类与细菌间，这些相似性都可以被清晰地探测到。基于以下理论，即生物都源自一个共同的祖先，它们在进化的过程中彼此产生联系，我们可以对这些相似性进行简明而自然的解释。人类本身与猩猩有着显而易见的相似性，如图1A所示，包括内部特征，例如我们的大脑结构与组成的

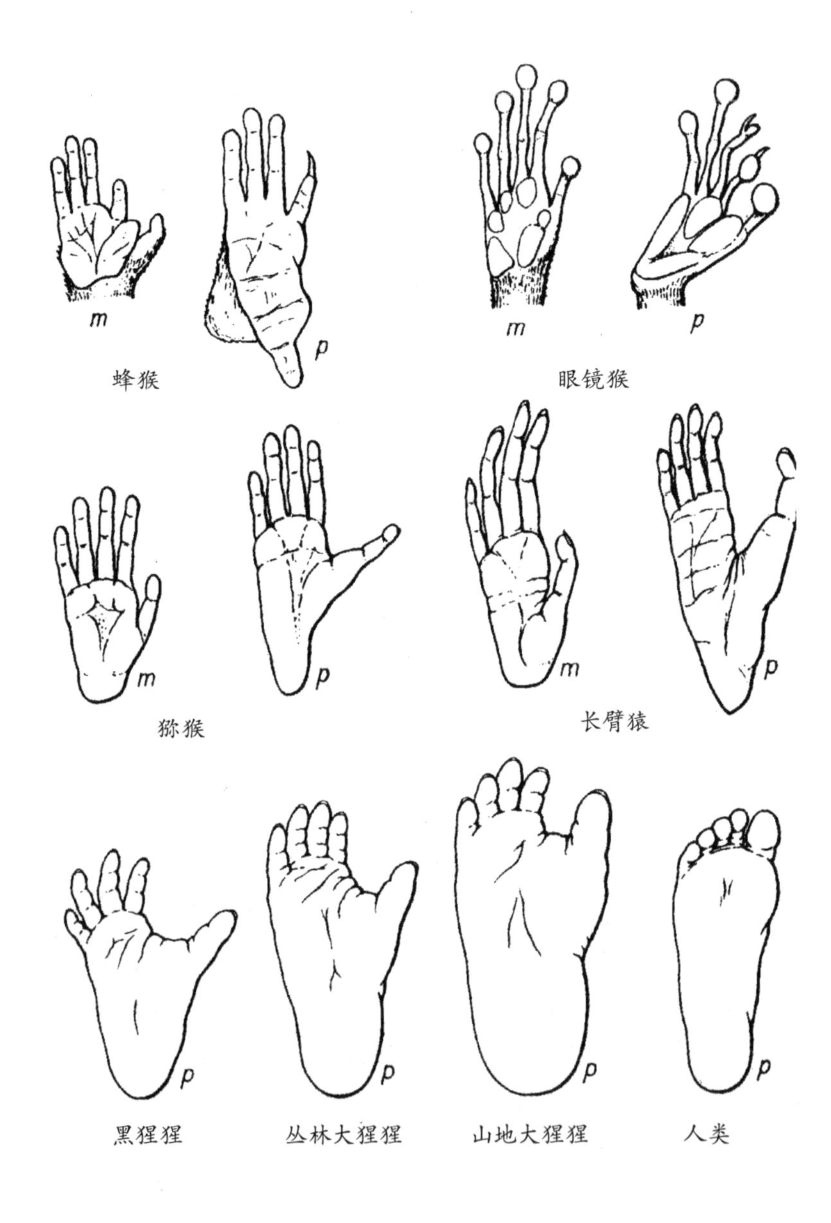


图1 A. 一些灵长类动物的手（*m*）与脚（*p*），展示了不同物种间的相似性，以及与动物生活方式相关的差异，例如树栖的物种有着与其他趾相对的趾（蜂猴和眼镜猴是原始树栖类灵长动物）。

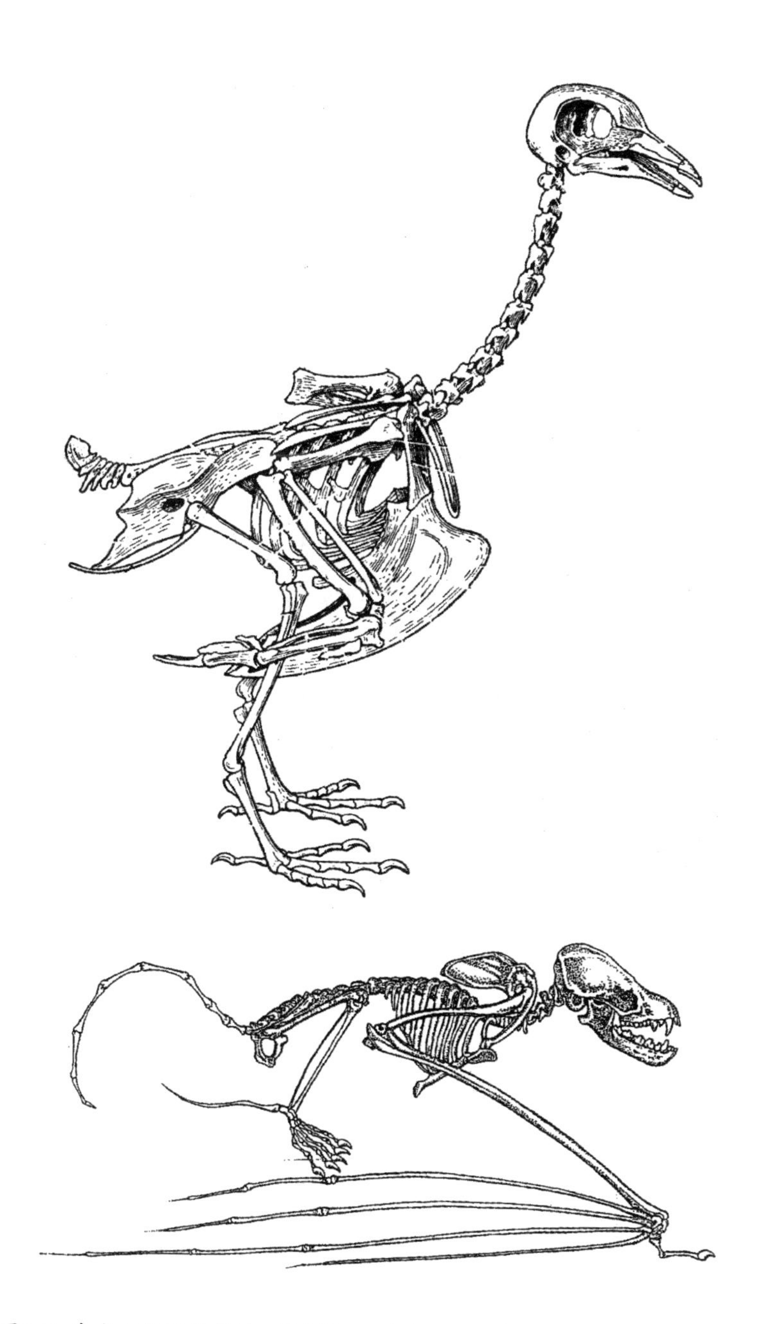

图1 B. 鸟类与蝙蝠的骨骼，图中展示了它们之间的相似与差异。

相似性。我们与猴子之间存在较小一些的相似性，甚至与其他哺乳动物间，尽管我们之间有那么多不同，也存在更小、不过依然十分明确的相似性。哺乳动物与其他脊椎动物相比，也存在许多相似之处，包括它们骨骼的基本特征，以及它们的消化、循环和神经系统。更让人惊奇的是我们与一些生物，例如昆虫之间存在的相似性（比如昆虫分节的躯干、它们对于睡眠的需求、它们睡眠与苏醒的日常节律的控制），以及不同物种间神经系统作用的根本相似性。

生物分类系统长久以来都基于易于观察的结构特点。例如，早在生物科学研究开始之前，昆虫就被认为是一类相似的生物；它们拥有分节的躯干、六对多节的足、坚固的外在保护壳等，这些使得它们与其他种类的无脊椎动物（例如软体动物）有着显著的区别。这其中的许多特征也存在于其他种类的动物身上，例如螃蟹和蜘蛛，只不过它们拥有不同数量的足（对于蜘蛛而言，这个数量是八条）。这些不同的物种都被归入同一个更大的分类之中，即节肢动物。昆虫是节肢动物的一类，而在昆虫之中，蝇类又组成了一小类，特征就是它们都只有一对翅，同时还有其他共有的特征。蝴蝶与蛾子形成了另外一个昆虫类别，这一类中的成员们两对翅上都有着精细的结构。在蝇类之中，我们依据共有的特征，将家蝇及它们的近亲与其他成员区分开来；在它们之中，我们又命名单个**物种**，例如最常见的家蝇。物种究其本质而言，即一群相似的能够彼此杂交繁殖的个体的集合。相似的种被划归进同一个**属**，同样地，同一个属中的生物都拥有一系列其他属所不具有的特性。生物学家通过两个名字确定每一个可鉴别的物种——属名，然后是该物种本身的种名，例如智人（*Homo sapiens*）；这些名字依据惯例采用斜体书写。

生物可以被逐级归入不同类别，随着归类的细化它们之间所共有的且其他类别的生物不具有的特征也越来越多——这一发现是生物学上的一个重大的进步。不同生物物种的划分，以及物种的命名体系，在达尔文之前很久就出现了。在生物学家开始思考物种的进化问题之前，对物种有一个清晰而具象的概念显然是非常重要的。对于生物这种分层次的相似性最简单也最自然的解释即，生物随着时间推移不断进化，自原始祖先开始不断多样化，形成了今天现存的生物类群，以及数不胜数的已灭绝生物（见第四章）。如我们将在第六章讨论的，如今可以通过直接研究它们遗传物质中的信息，对生物类群间这种推测的谱系关系进行判断。

另外一系列能够强有力地支持进化论的事实是：在不同物种中，存在同一结构的不同特化（modification）。例如，蝙蝠与鸟类翅膀的骨骼清晰地说明它们都属于特化的前肢，尽管它们与其他脊椎动物的前肢看起来截然不同（图1B）。类似地，尽管鲸的上肢看起来非常像鱼类的鳍，同时也显然非常适合游泳，它们的内在结构却与其他哺乳动物的足相似，除了趾的数目多了一些。结合其他证明鲸是哺乳动物的证据（例如它们用肺呼吸、给幼崽哺乳），这一事实也就合情合理。化石证据证明，陆生脊椎动物的前肢与后肢源自肉鳍鱼类的两对鳍（肉鳍鱼类中最著名的现存生物代表是腔棘鱼，见第四章）。而最早的陆生脊椎动物化石，也确实有多于五个的趾，就像鱼类与鲸。另一个例子是哺乳动物耳朵中的三块听小骨，它们负责把外界的声音传输给将声音转换为神经信号的器官。这三块微型的骨头最初发育自胚胎时期的下颌与颅骨，在爬行动物中它们随着发育逐渐扩大，最终形成头部与下颌骨骼的一部分。连接爬行动物与哺乳动物的化石纽带展示了这三块骨头在成年个体中连续的进化与变形，最终进化成为听小骨。在不同功能需求的作用下，相同的基本结构在进化过程中发生了显著的变化——类似的例子比比皆是，以上例子只是众多已知事例中的一小部分。

## 胚胎发育与痕迹器官

胚胎发育也为不同生物间的相似性提供了许多醒目的证据，清晰地显示了来自共同祖先的传承。不同物种的胚胎形成通常呈现惊人的相似，尽管它们的成熟个体千差万别。例如，在哺乳动物发育的某一阶段，会出现类似鱼类胚胎的鳃裂（图2）。如果我们是源自类似鱼类的祖先，那么这一切都有了很好的解释，否则这一现象将十分令人费解。正是由于成熟个体的结构需要使生物个体适应其所生存的环境，它们极有可能被自然选择所改造。可能发育中的血管需要鳃裂的存在，以引导它们在正确的部位形成，因此这些结构依然保留着，甚至在那些从不需要鳃功能的动物身上。然而，发育是能够进化的。在其他很多细节上，哺乳动物的发育与鱼类有着显著区别，因此其他在发育过程中影响不那么深远的胚胎结构，逐渐地消失了，取而代之的是新的结构。

相似性并不仅仅局限在胚胎阶段。**痕迹器官**长久以来也被认为是现代生物的远古祖先功能器官的残余结构。它们的进化非常有趣，因为这些实例告诉我们进化并不总是创造、改进结构，它们有时也会削减结构。人类的阑尾是一个典型的例子。作为消化道的一部分，阑尾在人体之中已经被大大缩减，而在猩猩身上，这部分依然巨大。在无腿动物身上出现退化的肢，这一例子也为人们所熟知。在发现的原始蛇化石之中，它们具有几乎

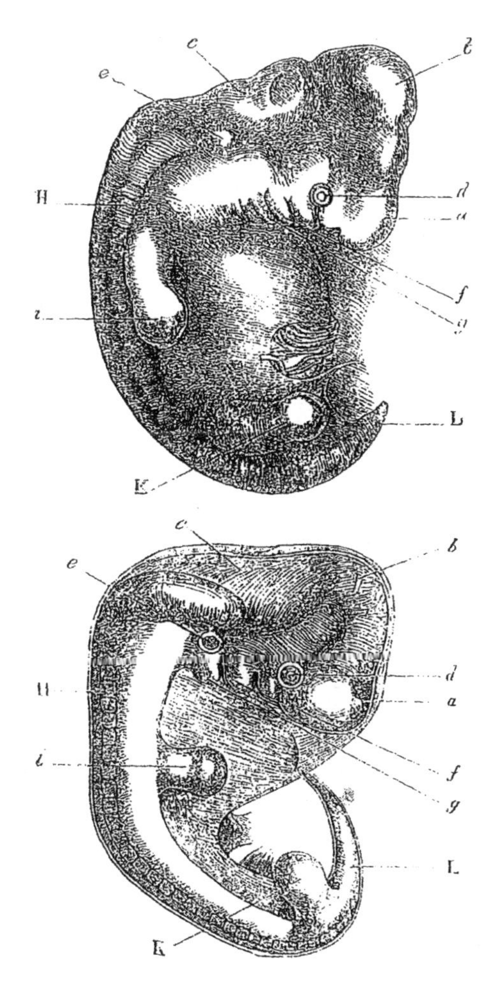


上图为人类胚胎（来自Ecker）；下图为狗的胚胎（来自Bischoff）。

a. 前脑、大脑半球等
b. 中脑、四叠体
c. 后脑、小脑、延骨髓
d. 眼
e. 耳
f. 第一腮弓
g. 第二腮弓
h. 发育中的脊椎和肌肉
i. 前肢
k. 后肢
l. 尾或尾骨

图2 人类与狗的胚胎，展现了它们在这一发育阶段最重要的相似性。在图中可以清楚看到鳃裂（标出了鳃弓f与g）。来自达尔文的《人类起源与性选择》(1871)。

完整的后肢，说明蛇是由有腿的、类似蜥蜴的祖先进化而来的。现代蛇的身体由一个瘦长的胸廓（胸部）以及众多脊椎骨（蟒蛇的脊椎骨超过300块）组成。对于蟒蛇而言，不带肋骨的脊椎骨标志着躯干与尾部的分界，也正是在这个部位发现了退化的后肢。在这个后肢中，有骨盆带与一对缩短了的股骨，它们的发育过程遵循了其他脊椎动物的正常轨迹，表达着通常控制四肢发育的相同基因。移植蟒蛇的后肢组织甚至能够促使鸡的翅膀形成额外的指，说明这部分后肢的发育系统依然存在于蟒蛇体内。然而，其他进化更加完全的种类的蛇，则彻底无肢。

## 细胞与细胞功能的相似性

不同生物间的相似性并不局限在可见的特征中。它们根深蒂固，深入到最细小的微观层面以及生命最基础的层面。一切动物、植物及真菌都具有一个基本特点，即它们的组织是由本质上相似的基本单元——**细胞**所组成的。细胞是所有生物体（病毒除外）的基础，从单细胞的细菌与酵母，到拥有高度分化组织的多细胞个体如哺乳动物。**真核生物**（所有非细菌的细胞生命）的细胞由**细胞质**与**细胞核**组成，在细胞核中包含了遗传物质（图3）。细胞质并不只是包裹在细胞膜内的供细胞核漂浮其中的简单液体，它含有一系列复杂的微小结构，其中包括许多亚细胞结构。其中两种最重要的**细胞器**是产生细胞能量的线粒体，以及绿色植物细胞中进行光合作用的叶绿体。现在，人们已经了解到，这两种细胞器都来源于侵入细胞并与细胞融合、成为其重要组成部分的细菌。细菌也是细胞（图3），但是相较而言，细菌细胞更简单，没有细胞核与细胞器。它们与和它们类似的生物被统称为**原核生物**。作为唯一一种非细胞形态的生物，病毒寄生于

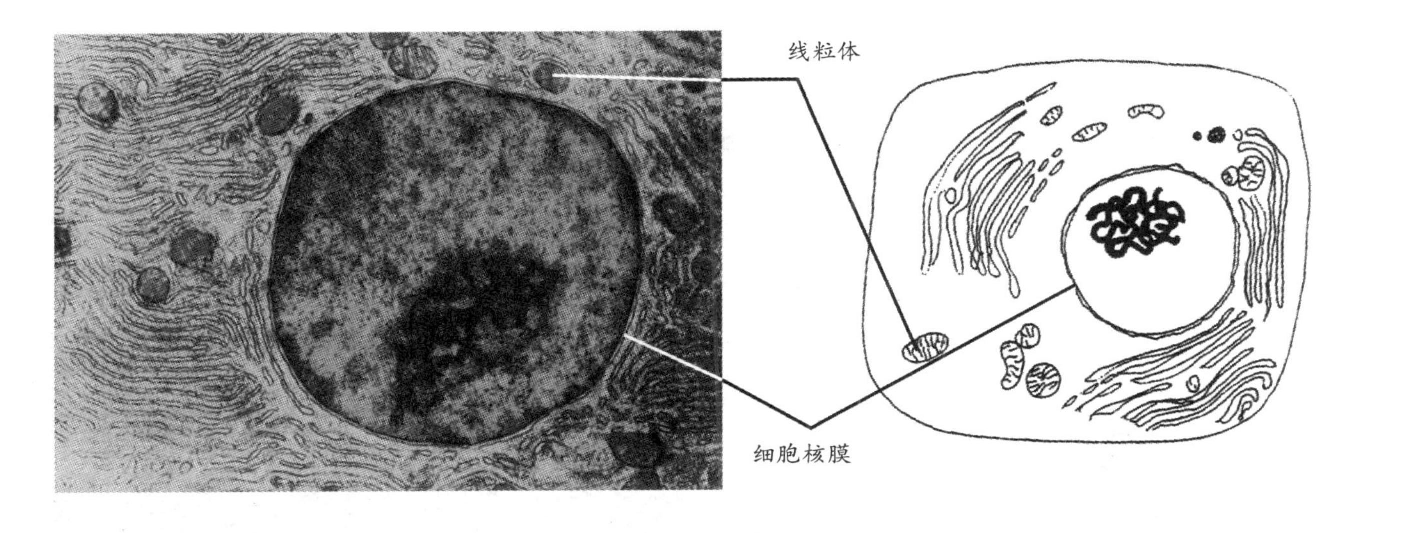


图3 真核生物与原核生物细胞。

A. 哺乳动物胰腺细胞电子显微镜照片与示意图，展示了核膜内包裹着染色体的细胞核，细胞核外的区域含有许多线粒体（这些细胞器也有包住它们的膜）；以及膜状的结构，它们参与蛋白质合成与输出，并将化学物质运入细胞。线粒体的体积略小于细菌细胞。

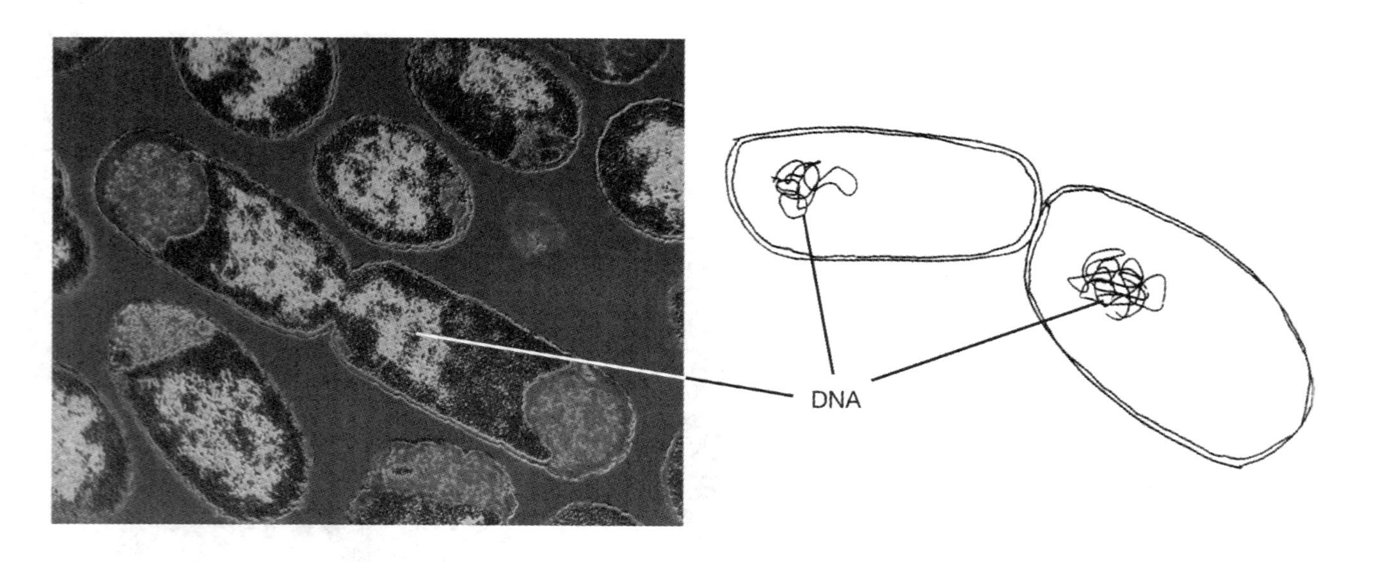


B. 细菌细胞的电镜照片及示意图，展现了它简单的结构：一层细胞壁与未包裹于细胞核中的DNA（脱氧核糖核酸）。

其他生物的细胞中进行繁殖。病毒仅由一个蛋白质衣壳及它所包裹的遗传物质组成。

细胞是非常微小而又高度复杂的工厂，它们生产生物体所需的化学物质、从食物原料中产生能量、形成生物结构（例如动物的骨头）。这些工厂里大部分的“机器”及许多的结构是**蛋白质**。一些蛋白质是**酶**，它们结合化学物质并在其上完成反应，比如像化学剪刀一样将一种化合物分解成为两种化合物。生物洗涤剂中的酶能够将蛋白质（如血迹或汗渍）分解为小片，从而使污渍能从脏衣服上被洗去；类似的酶存在于我们的消化道中，它们将食物中的大分子分解为更小的分子，从而能被细胞所吸收。生物体中的其他蛋白质还具有储存或运输的作用。红细胞中的血红蛋白能够运送氧气，肝脏中的一种被称为铁蛋白的蛋白质能够结合并储存铁元素。同时也存在结构蛋白，例如组成皮肤、毛发以及指甲的角蛋白。另外，细胞还产生向其他细胞或器官传递信息的蛋白质。激素是常见的交流蛋白，它们随着血液循环流动，控制众多的机体功能。另外一些蛋白质分布在细胞的表面，参与同其他细胞的交流。这些互动交流包括使用信号控制发育过程中的细胞行为、受精过程中卵子与精子间的交流以及免疫系统对寄生生物的识别等。

就像任何其他工厂一样，细胞受复杂机制的调控。它们对来自细胞外的信息进行响应（依靠横跨细胞膜的蛋白质，就像一个匹配外来分子的锁眼，见图4）。感觉感受器蛋白，例如嗅觉感受器及光学感受器，被用于细胞与环境间的信息交流。来自外界的化学信号与光学信号被转换成为电脉冲，沿着神经传导到大脑。迄今为止，所有已被研究过的动物在进行化学与光学感受时使用的蛋白质大体上都是相似的。为了说明在不同生物的细胞间发现的相似性，我们选取在苍蝇的眼睛与人类的耳朵中都存在的肌球蛋白（类似肌肉细胞中的蛋白质）为例，这种蛋白的基因突变会造成耳聋。

生化学家们已经将生物体中的酶划分成了许多不同类别，在一个全球性的编码系统中，每一种已知的酶（在像人类这种复杂动物体内，存在成千上万种酶类）都有一个编码。由于在种类极广的生物的细胞中存在如此多的酶类，因此这个系统对酶的

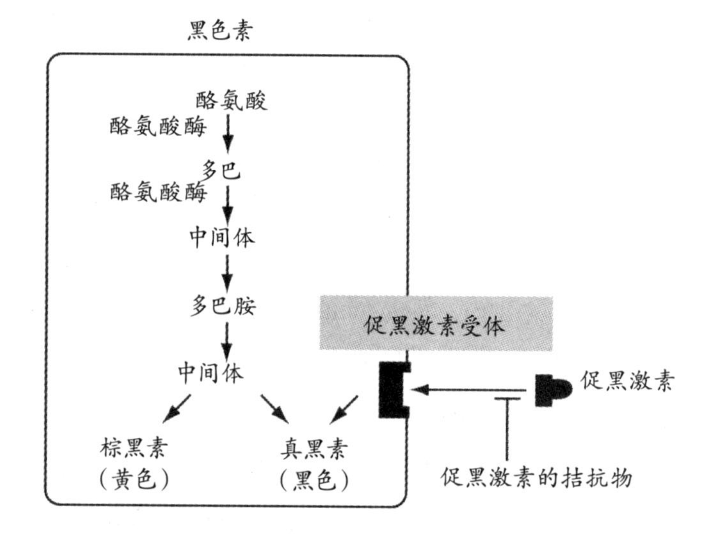


图4 在哺乳动物的黑色素细胞中，从氨基酸前体——酪氨酸合成黑色素与一种黄色素的生物途径。这个途径中的每一个步骤都由一种不同的酶催化。有效酪氨酸酶的缺失将会导致动物的白化病。促黑激素受体决定了黑色素与黄色素的相对量。这种激素的拮抗物的缺失将导致黑色素的形成，但这种拮抗物的存在“关闭”了该受体，导致了黄色素的形成。这就是虎斑猫以及棕毛鼠毛发中黄色与黑色部分形成的原因。让拮抗物失去作用的突变导致了更深的毛色；然而，黑色动物并不总是由这个作用产生，有些只是由于它们的受体一直保持在“开启”的状态，无论它们的激素水平是高还是低。

归类依据的是它们的功能而不是它们来源的生物。其中的一些，例如消化酶，负责将大分子分解成为小片段；一些其他的酶，负责将小分子聚合在一起；另外一些则负责氧化化学物质（将化学物质与氧气结合），等等。

将食物转化成为能量的方式在各种类型细胞中都大体相同。在此过程中，存在一个能量的来源（在我们的细胞中是糖或脂肪，但对于某些细菌而言，是其他化合物，例如硫化氢）。细胞通过一系列化学步骤分解最初的化合物，其中的某些步骤释放出能量。这种**代谢途径**就如同一条流水线，包含一连串的子流程。每个子流程都由它们自己的蛋白质“机器”完成，这些蛋白质“机器”就是这个代谢过程中不同步骤所对应的酶。相同的代谢途径在许多生物中都产生作用，在现代的生物学课本中介绍那些重要的代谢途径时，不需要去指明某种具体的生物。例如，蜥蜴在奔跑后感到疲倦，这是由于乳酸的堆积产生的，就像我们的肌肉中出现的情形一样。除了从食物中产生能量，细胞中同样存在着生产许多不同化学物质的代谢途径。例如，一些细胞产生毛发，一些产生骨骼，一些产生色素，另外一些产生激素，等等。对于表皮的黑色素而言，不管是对于我们人类、其他哺乳动物，还是翅膀具有黑色素的蝴蝶，甚至是酵母（例如黑色孢子），其产生的代谢路径都是相同的，而这个代谢路径中所使用的酶也被植物用于生产木质素（它是木材的主要化学成分）。从进化角度思考，从细菌到哺乳动物，代谢途径基本特征的根本相似性再一次变得很容易理解与接受。

这些细胞及身体功能中的每一种不同蛋白质都是由生物基因中的一种决定的（我们将在本章后文中详细解释）。而酶是所有生化途径发生作用的基础。如果在代谢途径中任何一种酶失去作用，将不能产生最终产物，就像流水线上的一环发生问题，产品就无法生产一样。例如，白化病突变是由于一种产生黑色素所必需的酶缺失造成的（图4）。阻断生产途径中的某一环是调节细胞产物的有效方法，因此细胞中存在着抑制剂，用来进行这种调节，正如前文黑色素生产中调节的例子。在另一个例子中，组织中存在着形成凝血块的蛋白质，然而在呈溶解态时，只有在这种前体物质的一部分被切除之后才会形成血块。负责切除的酶同样存在于组织中，但通常呈休眠状态；当血管遭到破坏时，因子被释放出来改变这种凝血酶，因此它立刻被激活，导致了蛋白的凝结。

蛋白质是由几十至几百条**氨基酸**亚单元之链构成的大分子，这些氨基酸单链与相邻的氨基酸相连，形成了蛋白链（图5A）。每个氨基酸都是一个相当复杂的分子，拥有它们各自的化学特性与大小。在生物体的蛋白质中共使用了20种不同的氨基酸；特



图5 A. 肌红蛋白（一种与红细胞中的血红蛋白相似的肌肉蛋白）的三维结构，图中可见蛋白质长链中所包含的氨基酸，编号为1至150，以及蛋白质中含铁的血红素分子。血红素结合氧气或二氧化碳，而这个蛋白的作用就是运输这些气体分子。

B. DNA结构，DNA是大多数生物遗传物质的载体分子。它包含两条互补链，相互环绕呈螺旋状。每一条链的主干由脱氧核糖分子（S）构成，通过磷酸分子（P）彼此相连。每个脱氧核糖分子对应一种被称作核苷酸的分子，它们构成了遗传学字母表的“字母”。存在四种类型的核苷酸：腺嘌呤（A），鸟嘌呤（G），胞嘧啶（C），胸腺嘧啶（T）。正如在双螺旋结构中所见，一条链上每种特定的核苷酸与另一条链上对应的核苷酸互补。这种配对的原则是：A与T对应，而G则与C对应。在细胞分裂过程中当DNA进行复制时，双链解开，遵循着上述配对原则，一个互补的子链将从各自的母链中产生。由此，在母链中A与T配对的位置将在子链中也同样为T与A。

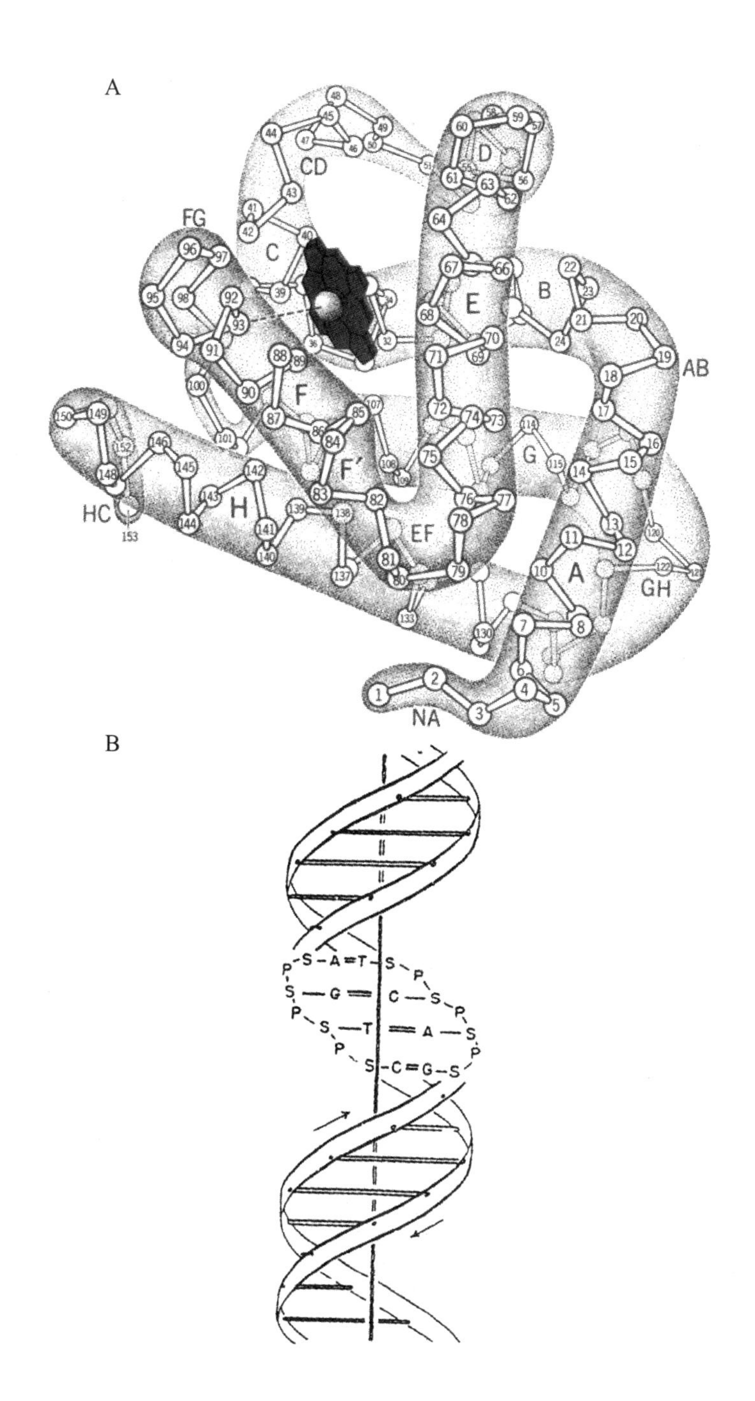
A
CD
D
FG
C
B
E
AB
F
F′
G
HC
H
EF
A
GH
NA
B
P-S-A=T-S-P
S-G=C-S-P
P-S-T=A-S-P
P-S-C=G-S

定的蛋白质，例如红细胞中的血红蛋白，具有一组有特定序列的氨基酸。一旦有了正确的氨基酸序列，蛋白质链就折叠成为功能蛋白的形状。蛋白质复杂的三维结构完全是由构成它们蛋白链的氨基酸序列决定的；而氨基酸序列则完全由生产此种蛋白的DNA的序列决定（图5B）——这一点我们即将详述。

对于不同的物种中相同的酶或蛋白质三维结构的研究表明，进化上差距极大的物种，例如细菌与哺乳动物间，尽管它们的氨基酸序列已经发生了巨大的改变，它们的蛋白结构经常存在相当大的相似性。我们在前文中提到的肌球蛋白就是一个例子，它在苍蝇的眼睛与哺乳动物的耳朵中都参与了信号传导。这种基本的相似性意味着（尽管十分令人惊讶），在酵母细胞中，往往可以通过引入具有相同功能的植物或动物基因而对代谢的缺陷进行纠正。通过细胞内一段人类基因的表达，具有由突变引起的铵盐摄取缺陷的酵母细胞被“治愈”了（这段基因用来编码Rh血型功能蛋白RhGA，该蛋白可能具有相应的功能）。野生型（未突变）酵母细胞中的这种蛋白与人类RhGA 蛋白在氨基酸上具有许多区别，然而在这个实验中，人类蛋白质却能够在缺少相应自体蛋白的酵母细胞中发挥作用。这个实验的结果也告诉我们，一个氨基酸序列被改变的蛋白质，有时也同样能够正常发挥作用。



## 生物共同的遗传基础

对于所有的真核生物（动物、植物，以及真菌）而言，遗传的物质基础在根本上是相似的。我们对于遗传机制的认识，是它通过某种物质（我们现在称作**基因**）对个体的许多不同性状进行控制。这一机制由格雷戈尔·孟德尔首先在豌豆中发现，但相同的遗传定律也适用于其他植物以及动物，包括人类身上。控制酶及其他蛋白产生（由此决定个体的特性）的基因是每个细胞中**染色体**所携带的DNA片段（图6，7）。人们最早在黑腹果蝇中发现

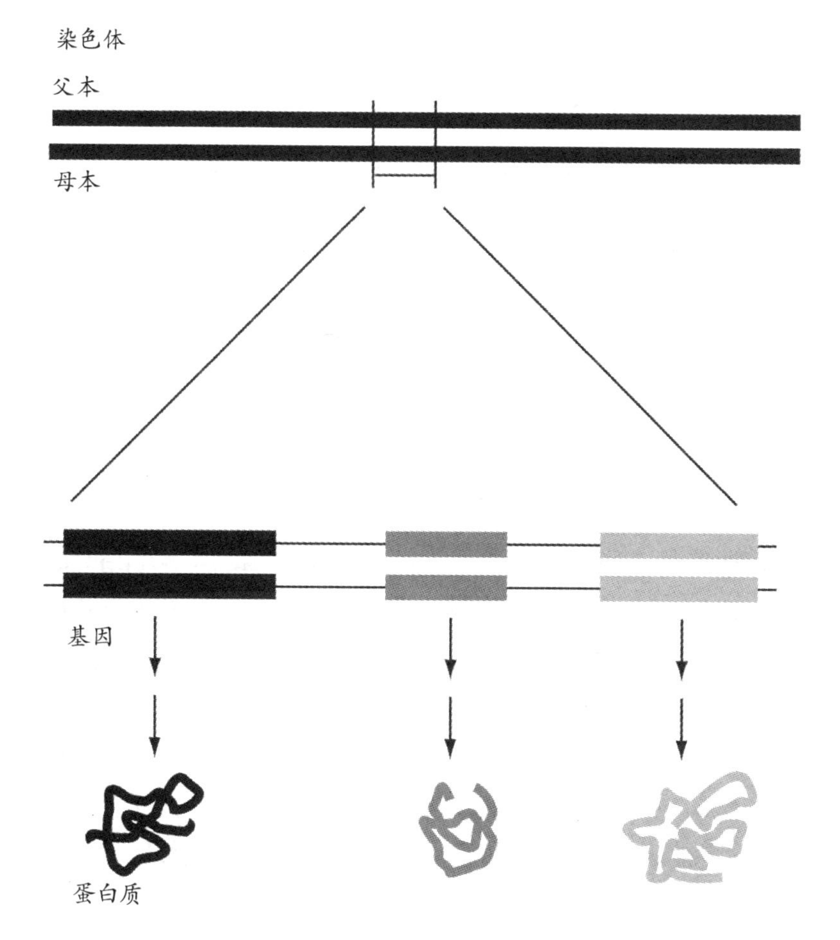


图6　一对染色体的图示，另有一个放大的示意图展示分布在该染色体上的三个基因，以及它们之间的非编码DNA。这三种不同的基因采用不同的灰度进行表示，表明每个基因都为不同的蛋白质编码。在实际的细胞中，这些蛋白中只有一部分能产生，其他的基因将被关闭，因此它们所编码的蛋白将不会形成。

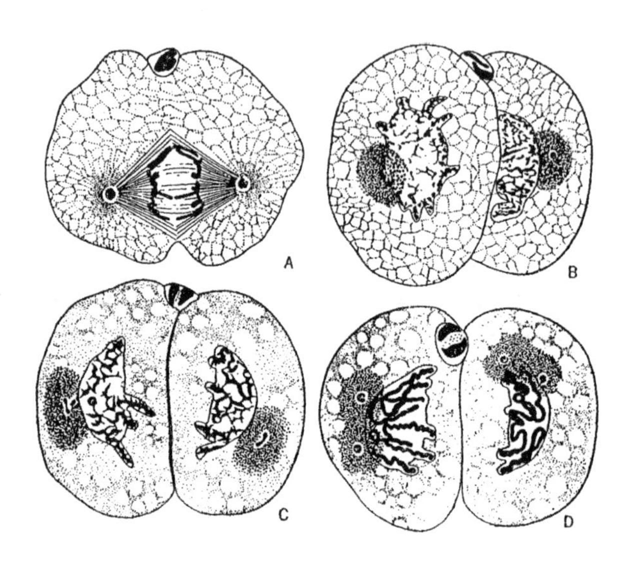


图7 一个正在分裂的线虫细胞，显示染色体不再被包裹在细胞核膜之中（A）；分裂过程中的不同时期（B，C）；以及最终形成的两个子细胞，每个子细胞都具有一个被核膜包裹的细胞核（D）。

基因在染色体上呈线性排列，这个规律同样适用于我们自身的基因组。染色体上的基因顺序在进化过程中可能会发生重新排列，不过这种变化发生的几率极小，因此我们可以在人类以及其他哺乳动物（例如猫、狗）的染色体上发现呈现相同顺序的相同基因。一条染色体本质上就是一个有着成千上万个基因的相当长的DNA分子。染色体DNA与蛋白质分子结合，这些蛋白质负责在细胞核中将DNA分子包裹成整齐的卷（类似整理电脑数据线的工具）。

在更高等的真核生物例如我们人类身上，每个细胞中都包含一套来源于母亲卵细胞核的染色体，以及一套来源于父亲精细胞核的染色体（图6）。在人类体内，父系或母系序列各有23条不同的染色体；在遗传学研究中常用的黑腹果蝇体内，染色体的数量是5（其中一对很小）。染色体携带有详细说明生物体蛋白质氨基酸序列所需的信息，同时还有决定什么蛋白将被生物体生产出来的调控DNA序列。

基因是什么？它如何决定一个蛋白质的结构？基因是**基因编码**的四个化学"字母"的排列，在它之中，三个相邻的字母（**三联体**）对应着该基因所编码的蛋白质中的一个氨基酸（图8）。基因序列被"翻译"成为蛋白质链上的序列；同时还有三联体标记氨基酸链的终点。基因序列上的改变导致了突变。这种改变中的大多数会使蛋白质在生产过程中出现一个不一样的氨基酸（但是，由于具有64种可能的DNA三联体字母，却只有20种氨基酸被用于蛋白质生产，有些突变将不会改变蛋白质序列）。纵观整个地球上的生物体，它们的基因编码差异非常微小，这充分说明了地球上的所有生物可能有一个共同的祖先。基因编码首先在细菌与病毒中进行研究，但很快就在人类身上验证并发现了共通性。在人类血红蛋白中，这个编码可能引起的几乎所有突变都已被人类检测到，但从来没有检测到对此特定编码来说不可能实现的突变。

为了生产出蛋白质产物，基因的DNA序列首先需要复制出一条"信使"，由相关联的RNA（核糖核酸）分子构成，它的"字母"序列由一种复制酶从原基因的序列中复制过来。信使RNA与一种由蛋白质聚合物和其他RNA分子组成的精巧的细胞器共同作用，将RNA中携带的信息翻译出来并产生出该基因所指定的蛋白质。这个过程在所有的细胞中本质上都是相同的，尽管在真核细胞中这个过程发生在细胞质中，信使RNA必须首先从细胞核中来到翻译过程所发生的细胞区域。在染色体中，这些基因之间是不编码蛋白质的DNA片段，这些**非编码**DNA中的一部

| | | | | | | | | | | | | | | | | | | | | |
|---|---|---|---|---|---|---|---|---|---|---|---|---|---|---|---|---|---|---|---|---|
| 人 | aac | cag | aca | gga | gcc | cgg | tgc | ctg | gag | gtg | tcc | atc | tct | gac | ggg | ctc | ttc | ctc | agc | ctg |
| 蛋白质 | Asn | Glu | Thr | Gly | Ala | Arg | Cys | Leu | Glu | Val | Ser | Ile | Ser | Asp | Gly | Leu | Phe | Leu | Ser | Leu |
| 人 | aac | cag | aca | gga | gcc | cgg | tgc | ctg | gag | gtg | tcc | atc | tct | gac | ggg | ctc | ttc | ctc | agc | ctg |
| 黑猩猩 | aac | cag | aca | gga | gcc | cgg | tgc | ctg | gag | gtg | tcc | atc | CcT | gac | ggg | ctc | ttc | ctc | agc | ctg |
| | | | | | | | | | | | | | Pro | | | | | | | |
| 狗 | aac | cag | acC | ggG | Ccc | cgg | tgc | ctg | gag | gtg | tcc | att | CcC | Aac | ggg | ctG | ttc | ctc | agc | ctg |
| | | | * | * | Pro | | | | | | | | Pro | * | | * | | | | |
| 老鼠 | aac | cag | Tca | gAG | CcT | Tgg | tgc | ctg | TaT | gtg | tcc | atc | CcA | gaT | ggC | ctc | ttc | ctc | agc | ctA |
| | | | Ser | Glu | Pro | Trp | | | Tyr | | | | Pro | * | * | | | | | * |
| 猪 | aac | cag | acG | ggC | Ccc | cAg | tgc | ctg | gag | gtg | tcc | atT | CCC | gac | ggg | ctc | ttc | ctc | agc | ctg |
| | | | | | Pro | Gln | | | | | | * | Pro | | | | | | | |

| | | | | | | | | | | | | | | | | | | | | |
|---|---|---|---|---|---|---|---|---|---|---|---|---|---|---|---|---|---|---|---|---|
| 人 | ggg | ctg | gtg | agc | ttg | gtg | gag | aac | gcg | ctg | gtg | gtg | gcc | acc | atc | gcc | aag | aac | cgg | aac |
| 蛋白质 | Leu | Gly | Leu | Val | Ser | Leu | Val | Glu | Asn | Ala | Leu | Val | Val | Ala | Thr | Ile | Ala | Lys | Asn | Arg |
| 人 | ggg | ctg | gtg | agc | ttg | gtg | gag | aac | gCg | ctg | gtg | gtg | gcc | acc | atc | gcc | aag | aac | cgg | aac |
| 黑猩猩 | ggg | ctg | gtg | agc | ttg | gtg | gag | aac | ATg | ctg | gtg | gtg | gcc | acc | atc | gcc | aag | aac | cgg | aac |
| | | | | | | | | | Met | | | | | | | | | | | |
| 狗 | ggg | ctg | gtg | agc | Gtt | gtg | gaa | aaT | gTg | ctg | gtg | gtg | gcc | Gcc | atT | gcc | aag | aac | cgC | aa |
| | | | | | * | | | * | Val | | | | | * | * | | | | * | |
| 老鼠 | ggg | ctg | gtg | agt | Ctg | gtg | gag | aaT | gTg | ctg | gtT | gtg | ATA | Gcc | atc | Acc | aaa | aac | cgC | aac |
| | | | | * | * | | | * | Val | | * | | * | * | | | | | * | |
| 猪 | ggg | ctg | gtg | agC | ctC | gtg | gag | aac | gTg | ctg | gtg | gtg | gcc | Gcc | atc | gcc | aag | aac | cgC | aac |
| | | | | | | | | | Val | | | | | * | | | | | * | |

图8 人类及部分哺乳动物中促黑激素受体蛋白（见图4）部分基因的DNA与蛋白质序列。图中仅显示了该蛋白中全部951个氨基酸中的40个。人类DNA序列在最上方，每三个DNA字母之间有空格，蛋白序列用灰色标记在下方（用三字母的代码表示不同氨基酸）。其他的物种见下排。与人类基因不同的DNA序列用大写字母标记。与人类DNA序列存在不同的三联体密码，如果它们为相同的氨基酸编码，则用星号标出；而编码与人类蛋白质序列不同的氨基酸的三联体则突出标示。许多拥有红色头发的人在三联体151位置上存在氨基酸的变化。

分具有重要的作用，它们是结合蛋白的结合位点，而这些结合蛋白将根据需要开启或关闭信使RNA的产生。例如，为血红蛋白编码的基因在发育成为红细胞的细胞中被开启，而在大脑细胞中则被关闭。

尽管不同生物的生活方式存在巨大差异——从单细胞生物到由亿万个细胞组成、具有高度分化组织的机体，真核细胞中所进行的细胞分裂过程都是类似的。单细胞生物，例如变形虫或酵母菌通过分裂出两个子细胞而进行繁殖。由卵子与精子融合而成的多细胞生物的受精卵，同样分裂出两个子细胞（图7）。而后发生了更多轮的细胞分裂，产生各类细胞与组织，构成成熟生物体。在一个成年的哺乳动物体内，有300多种不同类型的细胞。每一种类型的细胞都有它特有的结构，产生特定类型的蛋白质。这些细胞在发育过程中于组织与器官中的分布与排列，是一项需要对发育胚胎细胞间相互作用进行精密控制的工作。基因或被开启或被关闭，以保证对的细胞在对的地点、对的时间产生出来。在某些被透彻研究的生物，例如黑腹果蝇中，我们已充分了解到这些相互作用如何使得果蝇从看起来无差别的受精卵最终发育成为复杂的躯体。人们发现，许多特定组织（例如神经）发育与分化过程中的信号传导过程，在所有的多细胞动物中都普遍存在着；而陆生植物则使用一套截然不同的体系，也许正如化石记录所显示的那样，多细胞动物与植物有着不同的进化源头（见第四章）。

当细胞分裂时，染色体中的DNA首先会复制，因此每个染色体都有两份。细胞分裂过程被精密控制，以确保对产生的DNA序列进行细致“校对”。在细胞中含有某些酶，它们依靠DNA复制方式中的某些特性，将新产生的DNA与旧的“模板”DNA区分开来。这使得复制过程中发生的大多数错误能够被检测和纠正，保证了细胞在进入下一个阶段——细胞自身的分裂过程前，模板DNA被完全准确地进行复制。细胞分裂的机制保证了每个子细胞都接收到了与母细胞完全相同的一套染色体（图7）。

大多数原核生物的基因（包括许多病毒的基因）同样也是DNA序列，它们的组成与真核生物染色体中的DNA有一些微小的区别。许多细菌的遗传物质只是一个环状DNA分子。然而，一些病毒，例如引起感冒以及艾滋病的病毒，它们的基因由RNA构成。DNA复制过程中的校对工作不会在RNA复制过程中进行，因此这些病毒的突变率异常之高，它们可以在宿主的体内飞速进化。就像我们将在第五章描述的，这意味着研发针对它们的疫苗的难度很大。

真核动物与原核动物在非编码DNA数量上差异巨大。大肠杆菌（一种生活在我们肠道中通常无害的细菌）拥有约4300个基因，其中能够为蛋白质序列编码的基因约占其全部DNA的86%。与之相对的是人类基因组中为蛋白质序列编码的基因占整个DNA总量的不到2%。其他的物种位于这两个极值之间。黑腹果蝇在全部1.2亿个DNA“字母”里拥有约1.4万个基因，约有20%的DNA由编码序列组成。我们还无法准确知道人类基因组中不同基因的数量。目前为止最精确的计数来自全基因组测序。它使得遗传学家们能够基于从已有基因研究中获得的信息，识别出可能为基因的序列。组成各类物种基因组，特别是人类自己的基因组（它的DNA数量是果蝇的25倍）的DNA浩如烟海，从其中发现这些序列是一项艰巨的任务。人类基因的数量大约是3.5万，比人们根据不同功能的细胞与组织种类所推测出的数量要小得多。一个人能够产生的蛋白质数量也许会远远大于这个数字，因为我们用来计数的方法不能检测到很小或非常规的基因（例如，包含在其他基因内部的基因，这种现象存在于许多生物中）。现在尚不清楚非编码DNA对于生物的生命有多么重要的作用。尽管其中很大一部分是由生活在染色体中的病毒及其他寄生体组成，但它们中的一部分拥有重要的作用。如上文所述，在基因之外存在一些DNA序列，它们可以与那些控制细胞中基因“开关”的蛋白质相结合。对基因活动的这种控制在多细胞生物中肯定具有比在细菌中远为重要的作用。

除了发现截然不同的生物都把DNA作为其遗传物质，现代生物学还揭示了真核生物生命周期中更为深刻的相似性，尽管也存在差异性——从单细胞的真菌如酵母菌，到一年生的动植物，再到长寿（尽管不是长生不老）的生物如我们人类及许多树木。许多真核生物（尽管不是所有）在每一代中都存在有性繁殖阶段，在这个阶段中，融合的卵子与精子中来自母亲与父亲的基因组（分别由$n$条不同染色体组成，这是我们所讨论的物种所特有的）相互结合，形成一个具有$2n$ 条染色体的个体。当动物产生新的卵子或精子时，这种$n$的情形通过一种特殊的细胞分裂方式又重新形成。在这种分裂方式中，每一对父本与母本的基因都排列好，在互相交换遗传物质形成父本与母本基因的嵌合体之后，染色体对彼此分开，与其他细胞分裂过程中新复制出的染色体彼此分开类似。在这个过程的最后，每个卵细胞核或精子细胞核中的染色体数目减半，但是每个卵或精子都有一套完整的生物基因。在受精过程中，当卵子与精子的细胞核融合时，又重新形成了二倍体。

有性生殖基本特征的进化一定远远早于多细胞动植物的进化，后者是进化舞台上的新人。这一点从有性生殖的单细胞与多细胞生物的繁殖共同点上可以清楚看出，同时在酵母与哺乳动物这般差异巨大的物种间发现了相似的参与控制细胞分裂与染色体行为的蛋白质与基因。在大多数单细胞真核生物中，二倍体细胞由一对单倍体细胞融合产生，它们随即分裂产生带$n$个染色体的细胞，其过程与上述多细胞动物的生殖细胞形成过程相似。在植物中，染色体数量由$2n$变为$n$的过程在精子与卵子形成前发生，但还是涉及同样类型的特殊的细胞分裂。例如，在苔藓植物中，有一个很长的生命阶段是由单倍体形成苔藓植株，在这种植株之上，精子与卵子形成并完成受精，之后开始短暂的二倍体寄生阶段。

在某些多细胞生物中不存在这种两种性阶段并存的模式。在这种“无性生殖”的物种中，亲本产生子代并不用经历染色体数量在卵子产生过程中从$2n$变为$n$。然而，所有的多细胞无性生殖物种都具有清晰地源自有性生殖的祖先的印记。例如，普通蒲公英是无性生殖的，它们的种子不需要授粉就能够形成，而花粉对于大多数植物的繁殖来说是必须的。这对于像普通蒲公英这样的弱小物种来说是一个优势，它可以迅速产生大量的种子，家里有草坪的人们都能亲眼看到。其他的蒲公英物种通过个体间的正常交配进行繁殖，而普通蒲公英与这些蒲公英的亲缘关系如此之近，以至于普通蒲公英依然产生能够使有性生殖物种的花受精的花粉。

## 突变及其作用

尽管在细胞分裂的DNA复制阶段具有修正错误的校对机制，但是依然会产生错误，由此就产生了突变。如果突变导致蛋白质的氨基酸序列改变，这个蛋白质就有可能发生功能障碍；

例如，它有可能不能正确地折叠，因此就可能无法正常工作。如果这个蛋白质是一个酶，就有可能会导致这个酶所在的代谢途径效率降低，甚至完全停滞，就像上文中提到的白化病突变的例子一样。结构或交流蛋白的突变可能会损害细胞功能，或是影响生物体发育。人类的许多疾病都是由此类突变导致的。例如，控制细胞分裂的基因的突变增加了癌症的发病风险。正如前文所述，细胞有精密的控制系统来保证它们只在一切准备就绪时（对突变的校对必须完成、细胞不能有被感染或有其他损害的迹象，等等）才进行分裂。影响这类控制系统的突变将会导致不受控制的细胞分裂，以及细胞系的恶性增殖。幸运的是，细胞中一对基因同时突变的概率很小，而一对基因中只要还有一个未突变的基因，通常就足以对细胞功能进行纠正。而一个细胞系要成功癌变还需要其他的适应条件，因此恶性肿瘤并不常见。（肿瘤需要血液供应，而细胞的不正常特性也必须躲过人体的监测。）然而，了解细胞分裂及其控制依然是癌症研究的重要内容。不同的真核生物细胞中这一过程是如此相似，以至于2001年的诺贝尔医学奖授予了酵母细胞分裂的研究，该研究证明了与酵母细胞控制系统相关的一种基因在一些人类家族性癌症中也发生了突变。

导致癌症的突变非常罕见，大多数导致其他疾病的突变也是一样。在北欧人群中，最普遍的遗传病是囊性纤维症，即使在这种情况下，相应基因的未突变序列也占全部人口基因数量的98%以上。那些导致重要的酶或蛋白质缺失的突变可能会降低该个体的生存或繁殖概率，由此，导致酶功能障碍的基因序列在下一代中的比例就会降低，最终就会被种群所淘汰。自然选择最主要的角色就是保持大部分个体的蛋白质及其他酶类正常运转。我们在第五章中将再次考察这一观点。

有一种重要的突变，它使得一种蛋白质无法由它的基因足量产生。这种情况通常是由于基因的正常控制系统出现了问题，可能是当基因应该被打开时没有及时打开，导致产物数量有出入，也可能是在合成过程完成前就停止了蛋白质的生产。其他的突变可能不一定阻止酶的产生，但可能会让酶出现缺损，就像一条生产线上如果有一件必需的工具或机器出现问题，那么整条生产线都会受到阻碍甚至是停产。如果蛋白质中出现一个或几个氨基酸的缺失，那么这个蛋白质可能无法正确发挥功能；如果在蛋白链上某个特定位置出现了异样的氨基酸，哪怕其他位置都一切正常，也会出现同样的情况。当自然选择不再发挥其筛选作用时，这种导致功能缺失的突变也可能对进化产生贡献（参见第二章与第六章中选择中性突变的传播方法）。约65%的人类嗅觉受体基因是“退化基因”，它们不产生有活性的受体蛋白，因此我们比老鼠或狗的嗅觉功能要差得多（这并不让人惊讶，考虑到相较于我们，嗅觉在它们的日常生活与交流中更为重要）。

同一个物种中的正常个体之间同样具有许多差异。例如，对于人类而言，不同个体对于特定化学物质的味觉或嗅觉感知能力有所不同，对用作麻醉剂的某些化学物质的降解能力也不同。缺少降解某种麻醉剂的酶的个体可能对该物质有强烈反应，但是这种酶的缺乏对于其他方面则不会有什么影响。相似的情况也出现在对其他药物或者食物的降解上，这是人类多样性的一个重要方面，关于这些差异的研究对于经常使用烈性药物的现代医学而言十分必要。

葡萄糖—6—磷酸脱氢酶（细胞从葡萄糖中获得能量的起始步骤所用的一种酶）的突变部分说明了上述差异。完全缺失此基因的个体将无法存活，因为在细胞能量产生过程中会产生有毒副产物，而这个酶参与的过程正是控制这种副产物浓度水平的关键所在。在人类种群中，有至少34种不同的该蛋白正常变体，它们不但能健康地存活，而且还能够保护它们的机体不受疟原虫侵害。这些变体与最常见的蛋白质正常序列存在着一个或几个氨基酸的差异。其中的一些变体在非洲及地中海地区广泛分布，在一些患疟疾的人群中，变异个体频繁出现。然而，在人们吃下某种豆子，或使用了某种抗疟疾药物时，其中的一些变异将导致贫血。著名的ABO血型及其他种类的血型是人类种群中常规多样性的又一例证：这些血型的产生是由于控制着红细胞表面蛋白质序列的多样性。促黑激素受体蛋白的多样性对于黑色素产生至关重要（见图4），它能导致头发颜色差异。许多拥有红发的人，他们的这种蛋白中有一条氨基酸序列被改变。正如我们将在第五章中讨论的，基因的多样性是自然选择发挥作用产生进化改变的必不可少的原料。



## 生物分类、DNA与蛋白质序列

一组新的重要数据为生物体彼此间通过进化紧密相连提供了清晰的证据，这些证据来自它们DNA中的字母。现在我们可以通过DNA测序的化学过程“阅读”这些字母。300多年以来，通过对动物与植物的研究，基于表观性状的生物分类系统逐渐发展；在当今最新研究中，通过比对不同物种间DNA与蛋白质序列，这一系统获得了新的支持。通过测定DNA序列间的相似性，我们可以对物种间亲缘关系有一个客观的概念。这部分我们将在第六章中详述，现在我们只需要了解到，一个特定基因的DNA序列将与亲缘关系更相近的物种更为相似，而亲缘关系更远的物种的序列间差异则会更大（图8）。物种间差异的增加与两个被比较的序列分开的时间长度大致是成比例的。分子进化的这一特性使得进化生物学家们能够对那些化石资料无法确定的时间节点进行估算——用一种被称作**分子钟**的工具。例如，我们前文提到某一物种染色体上基因顺序发生改变，分子钟可以用来估算这种染色体重排的比率。与进化论观点一致的是，我们认为是近亲的物种，例如人类与猕猴，较之人类与新大陆的灵长类动物如绒毛猴，它们之间的染色体重排的差异更小。

在下一章中，我们将基于化石资料，根据现存物种的地理分布数据，对进化的证据进行阐释。这些观察结果补足了前面所述内容，表明进化理论为千姿百态的生物现象提供了一种自然的解释。

第四章

# 进化的证据：时空的印记

然而，人类的历史，不过是时间的长河中一道短暂的涟漪。

——摘自《论自然力的相互作用》

赫尔曼·冯·亥姆霍兹，1854

## 地球的年龄

18世纪末19世纪初，地质学家们成功地确认：地球现在的结构是长期不间断物理过程的产物；如果没有这一发现，人们不可能意识到生命由进化产生。其所使用的方法在本质上与历史学家和考古学家们所使用的方法相类似。正如伟大的法国博物学家布丰伯爵在1774年所写的那样：

> 正如在文明史研究中，我们查阅资料、研究徽章、破译前人的铭文，以便考证人类革命的新纪元、确定道德事件的发生时间，在自然界的历史中，我们也必须对整个地球的资料进行深入挖掘，从地球深处掘取古老的遗迹，把它们的碎片拼凑到一起，把这些物理变化的痕迹重新组合成为一个完整的证据，这个证据能让我们回到自然界的不同时代。这是在这个广袤无垠的空间里确定一个时间点、在不朽的时光岁月里树立一座里程碑唯一的方式。

尽管有把问题过度简单化的风险，两种关键的见解依然为早期地质学带来了成功：**均变论**原则，以及采用**地层学**划分年代。均变论与18世纪后期爱丁堡的地质学家詹姆斯·赫顿有着紧密的联系，并在之后由另一位苏格兰科学家查尔斯·赖尔在他的著作《地质学原理》（1830）中系统成文。该理论只不过是将天文学家用来理解遥远的恒星与行星的原理应用于地球构造的历史中，即其中所涉及的基本物理过程在任何时间、任何地点，都被认为是相同的。随时间推移而发生的地质变化反映了物理规律的作用结果，而物理规律本身是不变的。例如，物理定律表明，太阳与月亮的引力作用造成的潮汐所带来的摩擦力，必定使地球的转速在数百万年间减慢了。现在一天的时间比地球最初产生时一天的时间要长得多，但引力的大小并没有变化。

当然，并没有独立的证据证明这种均匀性的假设，就像没有任何有逻辑性的证据支持自然界具有规律性的设想，而这一设想正是我们日常生活最基本层面的基础。事实上，这两种假说之间并不存在区别，只是它们所应用的时间与空间尺度不同。它们的支持证据是：首先，均变论代表了可能的基础中最简单的一种，在此基础上我们能够对时间与空间上非常久远的事件进行诠释；其次，它已取得了令人瞩目的成功。

地质学上的均变论假说认为：火山活动及江河湖海的沉积物形成新的岩石，风、水流与冰的作用侵蚀古老的岩石，这些作用的累积结果在当今地球表面的构造中得到了体现。**沉积岩**（例如砂岩或石灰岩）的形成有赖于其他岩石的侵蚀。与之相对，火山作用或地震导致陆地上升形成山脉必定发生在岩石被侵蚀前。可以观察到的是，这些过程在今天依然在继续；去过山区的人们，特别是在一年之中冰雪冻结及消融时节去的人们，一定能观察到岩石的侵蚀作用，以及形成的碎片顺着河流被冲到下游。在河口，我们也很容易观察到堆积的沉积物。火山与地震活动局限在地球上某些特定的区域，特别是大陆的边缘和大洋的中心，其原因现在已广为人知，不过火山运动形成新的海岛、地震导致陆地上升的事件记录也为数不少。在《小猎犬号航海记》中，达尔文记录了1835年2月在智利发生的一次地震带来的后果：

> 这次地震最让人印象深刻的后果就是陆地的永久性上升；可能把它称作原因更为恰当些。毫无疑问康塞普西翁湾附近的土地上升了两三英尺……圣玛利亚岛（约30英里外）的上升更加显著；在其中一块地方，费兹洛伊船长在高水位线之上10英尺的地方发现了依然附着在岩石上的腐败的贻贝壳……这块区域的抬升特别有意思，它已经成为了一系列剧烈的地震集中的舞台，在它的陆地上，散布着数量巨大的贝壳，累积的高度达600甚至是1000英尺。

依照这些过程，地质学极为成功地解释了地表或地表附近区域地球的结构，同时重建了造成地球上诸多区域如今形态的地质事件。这些事件的先后顺序可以通过地层学的原理进行确定。人们用在不同岩层中发现的矿物成分与化石分布来描述不同岩层的特征。化石是早已死亡的植物和动物被保存下来的残骸而非矿物质形成的人工制品，这一认识是地层学获得成功的关键。在特定的沉积岩层中发现的化石种类能够提供它所形成的时代的环境信息。例如，我们通常可以分辨出该生物是海生、淡水生还是陆生。当然，在花岗岩或玄武岩这类由地壳以下熔融的物质所凝结成的岩石之中，并没有发现化石的踪迹。

19世纪早期，英国的河道工程师威廉·史密斯在走遍大不列颠修筑运河的过程中，发现在大不列颠岛上的不同区域存在相似的岩层变化（对面积如此小的土地来说，其不同时期的岩石种类异乎寻常地多）。基于旧岩层通常位于新岩层之下的原则，不同区域岩层演替的比较使得地质学家能够重建过去极为漫长的时间里岩层依次形成的顺序。如果在一个地点，A岩石位于B岩石之下，而在另一地点B岩石位于C岩石之下，我们可以推出顺序为A—B—C，即便A与C从未在同一个地点被发现。

19世纪的地质学家对这种手段的系统应用使得他们能够确定地质年代的大致分布（图9）。这种分布是一个相对而非绝对的年代表，要确定绝对的时间需要有方法对这个过程中所涉及的每一步的速率进行校正，而这样做是极为困难的，且不说精度如何。景观形成的过程十分缓慢，岩石的侵蚀每发生几毫米都需要许多年时间，沉积岩的形成也相应地十分缓慢。与之相似，即使在造山运动最活跃的区域，例如安第斯山脉，陆地上升的速率也不过是平均每年零点几米。在地球上的许多地方，由上述方式形成的沉积岩已经有数千公里的深度，且有证据表明，

图9 地质年表的大致划分。上表展示了寒武纪以来的各个被命名的时代，在这个时间段里，发现的化石数量最多（而它占地球年龄不到1/8）。[①]下表展示了地球历史上发生的重要事件。

---

① 表中第三纪最后给出的地质时代开始时间，数据与最新地质年代表有一定的出入。依据最新的地质学研究，第三纪现分为古近纪（含古新世、始新世和渐新世）和新近纪（含中新世和上新世）。——译注，下同

| 代 | 纪 | 世 | 距今 |
|---|---|---|---|
| 新生代 | 第四纪 | 全新世 | 1万年 |
| | | 更新世 | 200万年 |
| | 第三纪 | 上新世 | 700万年 |
| | | 中新世 | 2600万年 |
| | | 渐新世 | 3800万年 |
| | | 始新世 | 5400万年 |
| | | 古新世 | 6400万年 |
| 中生代 | 白垩纪 | | 1.36亿年 |
| | 侏罗纪 | | 1.90亿年 |
| | 三叠纪 | | 2.25亿年 |
| 古生代 | 二叠纪 | | 2.80亿年 |
| | 石炭纪 | | 3.45亿年 |
| | 泥盆纪 | | 4.10亿年 |
| | 志留纪 | | 4.40亿年 |
| | 奥陶纪 | | 5.30亿年 |
| | 寒武纪 | | 5.70亿年 |

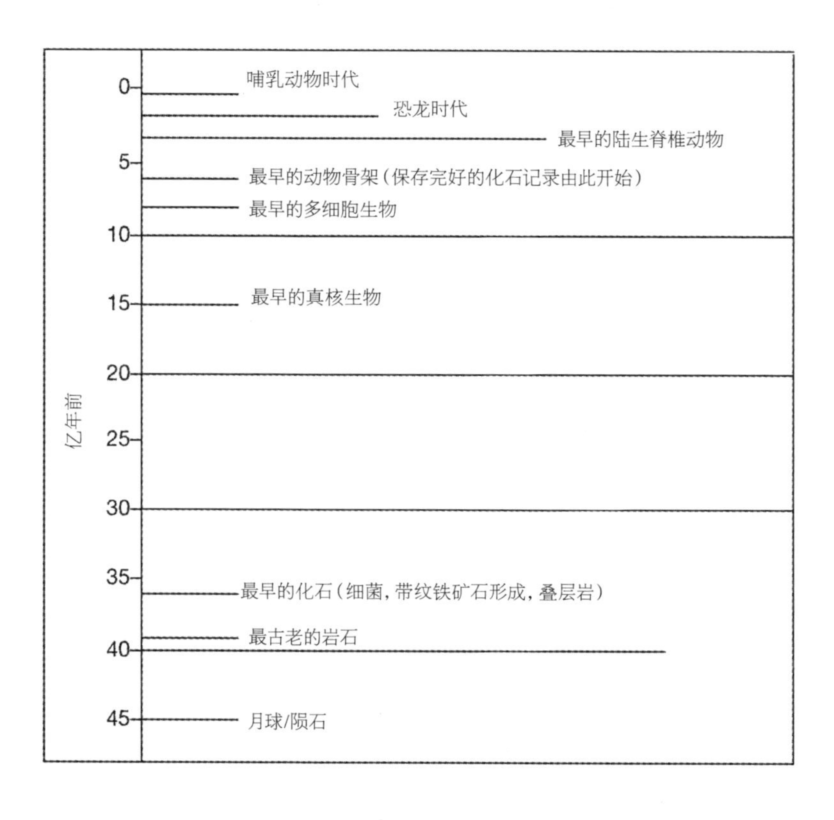

被侵蚀的沉积物也与之相去不远——鉴于这些，人们很快意识到，地球存在的时间至少得有数千万年，这与《圣经》所记载的年表是矛盾的。赖尔在此基础上提出：第三纪持续了约8000万年，而寒武纪则开始于2.4亿年之前。杰出的物理学家开尔文勋爵并不同意地球具有如此长的历史，他认为，如果地球真的已形成超过一亿年，那么最初那个熔融状态的地球的冷却速率将使得地球的中心比它实际上的温度要低很多。开尔文的计算在当时的物理学背景下是正确的。然而，在19世纪末，人们发现了不稳定的放射性元素——例如铀，能够衰变成为更为稳定的衍生物。这个衰变过程伴随着能量的释放，这些能量足以使得地球的冷却速率减慢，直至与它当前的预测年龄相符的数值。

放射性也为确定岩石样本的年代提供了全新而可靠的手段。放射性元素原子衰变成为更为稳定的子元素，并释放出辐射，这一速率是每年恒定的。当岩石产生时，可以假设其中我们所关注的元素是单一的；而后，当我们检测到样本中衰变所获得的子元素的比例，如果通过实验知晓衰变的速率，我们就能够估计这块岩石的形成时间。不同的元素可以用来测定不同时期的岩石。通过这一技术可以确定不同地质年代岩石的年代，这为我们提供了如今所公认的时间节点。尽管方法经常更新，而所确定的时间点也在不断地修正，但它们所预测的大致时间序列十分清晰（图9）。它为生物进化的发生，划定了一个广阔到不可思议的时间范围。



## 化石记录

化石记录是生命历史留给我们的唯一直接的信息来源。为了正确地对其进行诠释，我们需要了解化石是如何形成的，以及科学家们如何对化石进行研究。在植物、动物或者微生物死亡后，它们的柔软部分几乎一定会迅速降解。只有在某些特殊的环境中，例如沙漠干燥的空气中或是琥珀具有保护作用的化学物质里，负责降解的微生物才不能对这些软组织进行分解。人们发现了许多值得关注的保存软组织的例子，有些甚至可以追溯到几千万年之前，例如被困在琥珀中的昆虫。但是，这些与其说是规律不如说是例外。甚至连骨架结构，例如昆虫与蜘蛛体外覆盖的坚硬几丁质，或是脊椎动物的骨骼与牙齿，最后都会被降解。不过，它们降解的速率相对更慢一些，这让矿物质有机会渗透其中，最终取代其中的有机物（这种现象有时也发生在软组织之中）。若非如此，它们也许会形成一个具有它们的轮廓、被沉积的矿物质包围的空壳。

化石最有可能在水生环境中形成。在江河湖海的底层，矿物质沉淀、沉积物形成。尽管对于某个特定个体，形成化石的几率非常小，但沉到底层的残骸仍有机会变为化石。因此化石记录的结果存在非常大的偏差：生活在浅海的海洋生物，由于沉积物不断形成，其化石记录是最好的，而飞行生物的化石记录则最糟糕。此外，沉积物的形成可能会被打断，例如气候变化或者海底抬升。对于许多类型的生物，我们几乎没有它们的化石记录；而对于其他一些生物，化石记录曾经中断过许多次。

对于这种被中断的不完整性带来的问题，腔棘鱼是一个很好的例证。这是一种拥有分裂鱼鳍的硬骨鱼类，它的祖先是最早登上陆地的脊椎动物。腔棘鱼在泥盆纪时期（4亿年前）曾大量存在，但是随后就逐渐减少。距今最近的腔棘鱼化石要追溯到约6500万年前，很长时间以来，人们都认为这类生物已经灭绝了。直到1939年，非洲东南海上科摩罗群岛的渔民捕获了一只长相怪异的鱼，最后人们发现它就是腔棘鱼。于是随后科学家们能够对活腔棘鱼的习性进行研究；而在印度尼西亚，人们又发现了一个新的腔棘鱼群。腔棘鱼在一段极为漫长的时间里一定都存在着，但是并没有留下任何化石证据，因为它们的数量很少，而且生活在海洋的深处。

化石记录的中断意味着人们很难找到一系列长时间不间断的生物遗迹，以此展现进化的假说所需要的或多或少的连续变化。在大多数的例子中，新种类的动植物在化石中第一次出现时都没有表现出与它们早期形态存在任何显著的关联。最著名的例子是“寒武纪大爆发”：大部分重要类别的动物，作为化石首次出现都集中在寒武纪时期，即5.5亿至5亿年前（这部分将在第七章中再次讨论到）。



不过，正如达尔文在《物种起源》中所坚决主张的，化石记录的基本特性为进化提供了有力的证据。自达尔文的时代以来，古生物学家们的发现一次又一次地巩固了他的论述。首先，人们发现了许多过渡物种的实例，这些物种将原先被认为中间有着不可逾越鸿沟的物种连接起来。始祖鸟也许是其中最著名的生物，在《物种起源》一书出版后不久，人们发现了这种既像鸟又像爬行动物的物种的化石。始祖鸟化石非常罕见（现存只有六个样品）。它们来自于约1.2亿年[①]前侏罗纪时期的石灰岩，这种岩石沉睡在德国一个大湖湖底。这些生物有着被拼接起来的特征，有些特征像现代的鸟类，例如羽毛与翅膀，而有些又像爬行动物，例如长着牙齿的颚（而不是像鸟一样的喙），以及长长的尾巴。它们的骨架结构中很多细节都与同时期的恐龙极为相似，

① 数据有误，应为1.5亿年。

但是始祖鸟很明显会飞，这一点又与恐龙有所不同。随后，人们又发现了其他将恐龙与鸟类联系起来的化石，最近人们又发现了在始祖鸟之前还存在过长着羽毛的恐龙。其他重要的中间类型包括来自始新世（约6000万年[①]前）的哺乳动物化石，这些动物拥有前肢及简化的后肢以适应游泳。它们连接了现代的鲸类与偶蹄目食草动物例如牛和羊。

随着人们取得越来越多的研究成果，许多化石记录的间断被填平了，对人类的研究就是一个很好的例证。在1871年达尔文关于人类进化的著作《人类起源》第一次出版时，人们尚未发现任何人类与猿之间相联系的化石证据。达尔文基于解剖学上的相似性，认为人类与大猩猩及黑猩猩之间的关系最为紧密，因此人类可能起源于非洲的祖先，而这些祖先同样也进化成为如今的猿类。在此之后，人们发现了一系列的化石证据，通过前文所述方法精准地确定了其年代，而后新的化石证据被持续发现。这些化石中，距今越近的化石与现代人类越相似（图10）。能被明显地归入智人物种的最早化石被确定产生于距今只有几十万年之前。与达尔文的推断相一致的是，早期人类的进化很可能发生在非洲，而我们的祖先们可能在约150万年前首次抵达了欧亚大陆。

在时间序列上几乎不间断的化石例证也同样存在，由此可以确定，我们能够发现在进化上呈现单一谱系变化的化石记录。对海底沉积物挖掘结果的研究是最好的例证，从这类挖掘之中，我们能够获得很长的岩层序列。这些岩石的主体由数不胜数的微生物化石组成，而上述研究使得我们能够精确判定这些微

① 数据有误，应为5000万年。

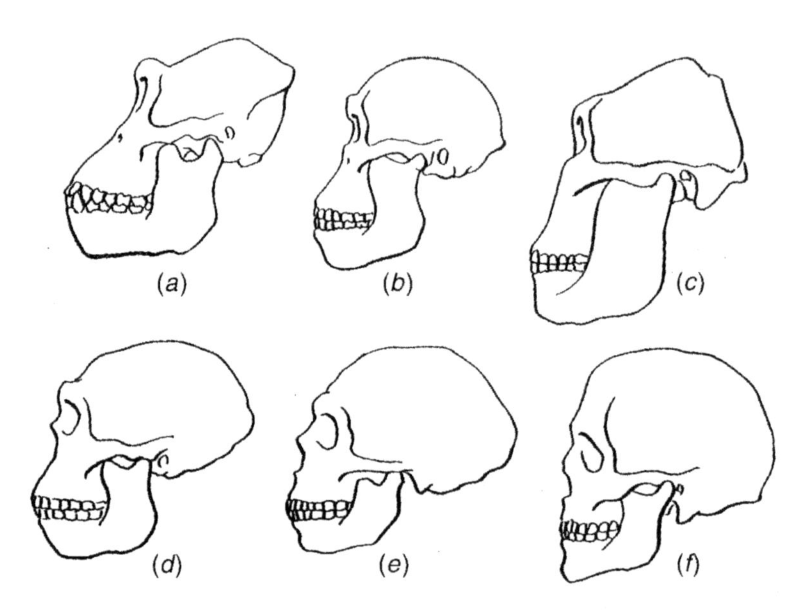


图10 某些人类祖先与近亲的头骨。(a)雌性大猩猩。(b)和(c)人类最早的近亲之一——南方古猿的两种不同化石(约300万年前)。(d)南方古猿与现代直立人之间的中间物种的化石。(e)尼安德特人的化石,距今约7万年。(f)现代人——智人。



生物连续样本的形成年代。对这些生物(例如有孔虫目,一种单细胞海洋动物)骨骼外形的仔细测定使我们能够描述,在一段漫长时间里渐次演替的种群在总体水平与变化程度方面的特征(图11)。

如果没有进化学说的支持,人们很难理解化石记录的基本特性,更不用说解释化石记录中过渡物种的存在。尽管寒武纪之前的化石记录非常不完整,但依然保存有超过35亿年前的细菌及与其相关的单细胞生物的遗骸。再经过很长一段时间,出现了更为高级的(真核生物)细胞的遗迹,但依然未发现多细胞生物出现的证据。由简单细胞群构成的生物直到约8亿年前才出

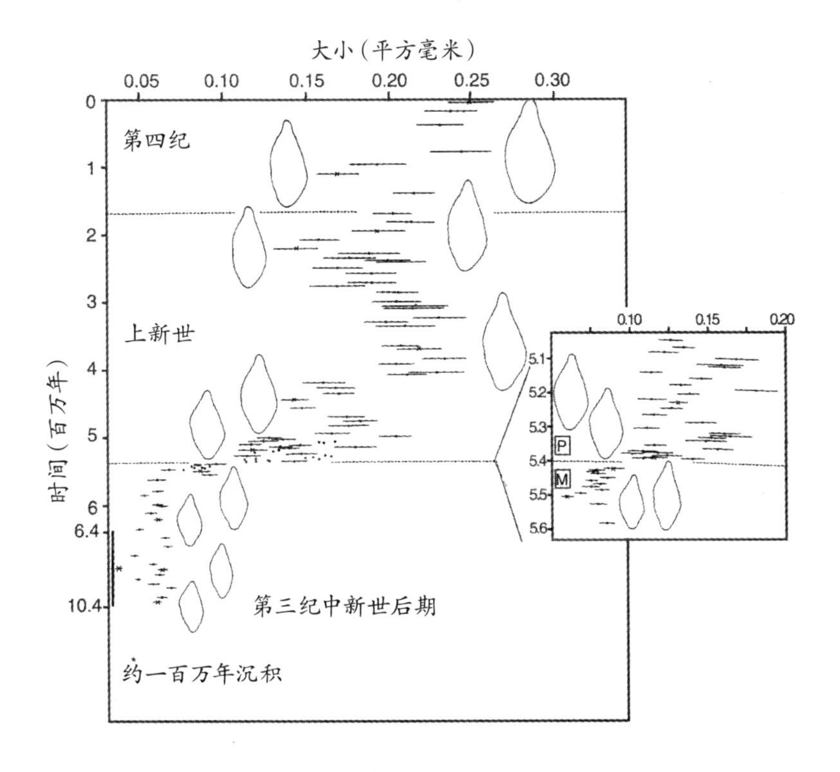


图11 化石中进化的渐变。此图展示了一种单细胞海洋甲壳动物——有孔虫化石样品中身体大小的均值与范围。在这个谱系之中，除了两个明显的间断点外，身体大小顺次变化。在第三纪中新世后期与上新世的交界点，一组更为确切的化石（见小图）说明了那些较为粗略的化石中所观察到的不连续几乎完全反映了一段剧烈变化时期，因为大部分连续的样本都彼此重叠。对于400万年前的间断点，迄今为止还未发现化石信息。

现，在那个时期环境极其恶劣，地球上的大部分被冰雪覆盖。约7亿至5.5亿年前，有证据表明出现了有着柔软身体的多细胞动物。

正如前文中提到的，拥有硬质骨骼的动物遗骸直到5.5亿年前的寒武纪岩石中才被大量发现。在约5亿年前的寒武纪末期，有证据表明几乎出现了所有主要的动物类群，包括原始的像鱼一样的脊椎动物，这种动物缺少上下颌，类似现代的七鳃鳗。

直到这个时期，所有的生物都与海洋沉积物有关，而藻类是唯一有遗骸的植物，它没有陆地多细胞植物输送液体所需要的导管。4.4亿年前，有证据表明出现了淡水生物，而后孢子化石的发现表明最早的陆生植物开始出现，类似鲨鱼的有颌鱼出现在海洋之中。在泥盆纪时期（4亿—3.6亿年前），淡水与陆生生物遗骸变得更为普遍与多样化。有证据表明原始的昆虫、蜘蛛、螨虫与多足动物开始出现，同样也出现了简单的维管束植物以及真菌。有颌的硬骨鱼类也逐渐变得普遍，其中就包括肉鳍鱼类，它与出现在泥盆纪末期类蝾螈的早期两栖动物结构相似。这些是最早的陆地脊椎动物。

在地质记录的下一个时期，石炭纪（3.6亿—2.8亿年前），陆地生物形态变得丰富且多样化。在热带沼泽之中生长的树状植物的遗骸化石形成了煤炭沉积，这也是这个时期命名的由来，这种植物更类似于同时期的杉叶藻与蕨类，与当代的针叶树或阔叶树没有关系。在泥盆纪的末期，原始爬行动物的遗骸出现，这是第一种完全脱离水的脊椎动物。在二叠纪时期（2.8亿—2.5亿年前），爬行动物出现了一个巨大的分化，有些的结构特点与哺乳动物日益相似（似哺乳爬行动物）。一些现代的昆虫类型，例如臭虫、甲壳虫，开始出现。

在化石记录中可以发现，二叠纪末期出现了最大规模的生物灭绝，一些之前占据优势地位的物种例如三叶虫突然完全消失，许多其他物种也几乎完全消失。在之后的恢复期，在海洋中与陆地上出现了许多新的物种。与现代的针叶树和阔叶树相似的植物在三叠纪时期（2.5亿—2亿年前）出现。恐龙、龟以及原始的鳄鱼出现了；就在三叠纪的末期，出现了最早的真正意义上的哺乳动物。与先驱们不同，它们的下颌含有一块与头盖骨直接相连的骨头（在爬行动物中组成这个连接的三块骨头进化成为哺乳动物耳朵里的三块听小骨，见第3章，第15页）。与现代鱼类相似的硬骨鱼类在海洋中出现了。在侏罗纪时期（2亿—1.4亿年前），哺乳动物开始或多或少地分化，但是陆地依然被爬行动物，特别是恐龙所统治。会飞的爬行动物与始祖鸟类出现了。苍蝇与白蚁首次出现，在海洋中出现了蟹类与龙虾。直到白垩纪时期（1.4亿—6500万年前），有花植物才进化出现——它们是主要生物中最晚进化出现的。现代主要的昆虫类型在这个时期都出现了。有袋类哺乳动物（有袋目）在白垩纪中期出现，与现代胎盘哺乳动物相似的类型在白垩纪末期也被发现。恐龙依然数量庞大，尽管在这个时期行将结束时出现了减少。

伴随着白垩纪的结束是史上最著名的物种灭绝事件，与一颗在墨西哥尤卡坦半岛着陆的小行星有关。所有的恐龙（除了鸟类）都消失了，一同消失的还有许多曾经在陆地与海洋中普遍存在的生物。接下来便是第三纪，它一直延续到大冰河世纪（约200万年前）的到来。在第三纪的第一阶段（6500万—3800万年前），胎盘哺乳动物的主要类型出现。最初，它们多半与现代的食虫类动物例如鼩鼱相似，但是在这个时代的末期，其中的一些变得相当独特（例如我们能够辨认出鲸与蝙蝠）。大多数的主要类别的鸟类与现代的无脊椎动物在这个时期出现，除了禾本科之外的所有主要有花植物也出现了。与现代种类基本一致的硬骨鱼数量增多。在3800万至2600万年前之间出现了草原，同样也出现了类马的食草动物，它们拥有三个趾头，而不是现代马的单个趾头。原始的猿类同样出现了。2600万至700万年前，在北美出现了大片的草原，拥有短侧脚趾与高冠齿、适应食草的马类出现。许多有蹄类动物，例如猪、鹿与骆驼，还有大象也开始出现。猿与猴子分化越来越大，尤其是在非洲。在700万至200万年前，海洋生物本质上较为接近现代，尽管其中很多物种如今已经灭绝。在此时期，出现了最早的具有明显人类特征的动物遗骸。在第三纪的末期（200万—1万年前），是一连串的冰河世纪。大多数的动植物基本具备现代形态。最后一个冰河世纪的末期（1万年前）至今，人类统治了陆地，许多大型哺乳动物开始灭绝。一些化石证据证明了这个时期的进化改变，例如在海岛上许多大型哺乳动物的矮型种的进化。

因此，化石记录表明生命起源于30亿年前的海洋，在10亿多年中，只存在着与细菌相关的单细胞生物。这正是进化模型所预期的；将基因编码转化成为蛋白质序列所需的装置，以及哪怕是最简单细胞的复杂结构，它们的进化必定需要许多步骤，其具体过程几乎超出了我们的想象。之后化石记录中出现的真核细胞的明显证据，以及它们总体上比原核细胞更为复杂的结构，与进化论也是相一致的。这同样适用于多细胞生物，从单个细胞发育而来的它们需要精密的信号传递机制以控制生长与分化，而这些在单细胞物种出现之前不可能进化出现。一旦简单的多细胞生物进化形成，可以理解的是它们会迅速分化成为各种形态，以适应不同生存类型，正如寒武纪所发生的那样。我们将在下一章讨论适应与分化。

从进化的角度来看，生命在相当长的一段时间里只属于海洋这个事实也变得很好理解。在地球历史的早期，有地质证据表明大气层中的氧气非常稀薄，于是缺少了由氧形成、能够阻挡紫外线的臭氧层，使得陆地甚至淡水中的生物很难存活。一旦早期细菌与藻类的光合作用导致氧含量变得充足，这一重障碍便消失了，于是生物开始有可能涉足陆地。有证据表明在寒武纪之前一段时间大气中的氧气含量出现了上升，这也许促进了更大且更多的复杂动物的进化。同理可知，会飞的昆虫与脊椎动物化石在陆地动物之后出现也是理所当然的，因为真正的飞行动物不大可能从纯粹的水生生物进化而来。

从进化学角度分析，生物类型周期性地增多与分化，之后又大规模地灭绝（三叶虫与恐龙的遭遇）或减少至一两个幸存种（如腔棘鱼类），这一现象同样合乎情理。进化的机制并没有前瞻性，也不能保证它们的产物能够从巨大而突然的环境变化中幸存。同理可知，在进入一个新的栖息地后（例如入侵陆地），或是在一个占据优势的竞争种灭绝之后（例如在恐龙灭绝之后的哺乳动物），物种的迅速分化也符合进化原则的预期。

因此，基于生物学知识对化石记录的解释符合地质学家应用于地球历史的均变论原则。化石证据也可能存在一些不符合进化论的实例。据说，伟大的进化学家与遗传学家霍尔丹在被问到什么样的观测结果会让他放弃对进化论的信念时曾回答道："一只寒武纪之前的兔子。"迄今为止，还未发现此类化石。

## 空间上的印记

正如达尔文在《物种起源》的十五章中花费两章内容所描述的，另一组只有基于进化论才能解释的重要事实来自生物的空间而非时间的分布。其中最惊人的例证之一是那些海岛（例如加拉帕戈斯群岛与夏威夷群岛）上的动植物们。有地质证据证明这些群岛是由火山运动形成的，它们从未与大陆相连。根据进化理论，这种群岛上现存的生物一定是那些能够穿越这些新近形成的海岛与最近的已居住海岛之间遥远距离的生物的后代。这对我们可能观测到的结果造成了一些限制。首先，外来物种要在新形成的岛屿上定居，其难度可想而知，这意味着很少有物种能够生存下来。其次，只有那些具备某些特征，能够穿越数万海里大洋的物种才能最终扎根。第三，即使在这些能够扎根的物种之中，也存在着许多不确定因素，因为能够到达岛上的物种数量极少。最后，在如此偏远的岛屿之上，进化所能形成的许多类型在其他地方都不可能出现。

这些设想都被很好地验证了。与有着相似气候的大陆或沿海岛屿相比，海岛上主要生物群的种类的确相对较少。在海岛上发现的生物种类（在人类进驻之前），是其他地区的非典型物种。例如，岛上通常具有爬行动物与鸟类，而陆地哺乳动物与两栖动物总是不存在。在新西兰，在人类进驻之前这里没有陆地哺乳动物，不过存在着两种蝙蝠。这说明蝙蝠能够横跨辽阔的海洋。在人类将许多物种引入之后，它们的疯狂蔓延说明当地的环境并非不适合它们生存。但即使是当地主要的动植物群，也经常出现整个群体消失的情况，而其他存活的物种也通常不成比例。因此，在加拉帕戈斯群岛上，陆地鸟类的种类只有20多种，其中14种是雀科鸟类，这些著名的雀类在达尔文搭乘小猎犬号环游世界时被记载在他的旅行手记中。这与地球上其他地区的情况不同，在其他区域雀类只是陆地鸟类中一个很小的组成部分。这正符合我们之前的预期：最初，只有很少种类的鸟类进驻这个岛屿，其中的一种为雀类，而它们成为了如今雀类物种的祖先。

正如这种观点所预期的，海岛的物种有着许多属于自身独一无二的特性，与此同时它们也表现出与大陆物种之间的联系。

例如，在加拉帕戈斯群岛上发现的植物种类中，有34%的物种从未在其他地区出现过。达尔文雀类的鸟喙大小与外形也比一般的鸟儿（通常拥有大而深的鸟喙，主要吃种子）要远为多样化，它们显然适应于不同的捕食模式（图12）。这些鸟喙中有些相当不同寻常，例如有着尖锐鸟喙的尖嘴地雀喜欢啄食筑巢海鸟的臀部，吸食它们的血液。鹀形树雀使用小树枝或仙人掌刺获取

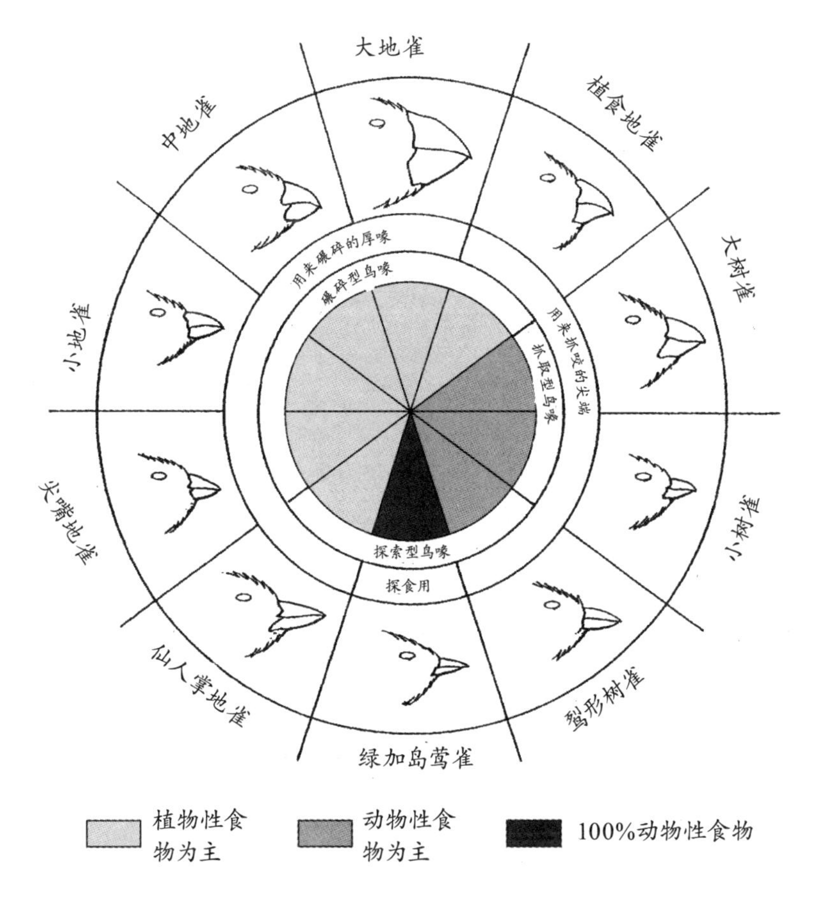


图12 达尔文雀类的鸟喙，展示了不同食性的物种的鸟喙在大小与形状上的差异。

枯木中的昆虫。更为壮观的疯狂进化例证来自海洋岛屿中的其他类群。例如，夏威夷岛上的果蝇种类数量超过世界上其他任何地方，而它们在身体大小、翅膀样式以及进食习惯方面差异巨大。

如果这些海岛物种的祖先首次进入此岛屿时，发现这个环境里没有已经到达的竞争者，这些观察结果就容易理解了。这种情况将会容许它们进化出与新的生活方式相适应的特性，使得原先的物种分化成为几种不同的后代。尽管在达尔文雀类中发现了许多不同寻常的结构与行为上的变异，但采用第三章与第六章的方法对它们的DNA进行的研究表明，这些物种在约230万年前有着共同的祖先，与大陆的物种亲缘关系也非常接近（图13）。

正如达尔文在《物种起源》一书中描述加拉帕戈斯群岛上的居住者时所写到的：

> 在这里，几乎所有的陆地与水生生物都有着来自美洲大陆的明确印记。这里有26种陆生鸟类，古尔德先生把其中的25种归为特殊种，它们应该是生于斯长于斯的；但是它们中的大多数与美洲物种在习性、姿态、叫声等各种特征上的相似都是显而易见的。其他的动物也是如此，还有几乎所有的植物——正如虎克博士在他有关这个群岛上植物的绝妙回忆录中所写的。这位博物学家看到这些距离大陆几百英里远、太平洋中的火山群岛上的生物时，依然觉得自己仿佛置身于美洲大陆上一般。为什么会这样呢？为什么这些本应该是加拉帕戈斯群岛上独创、别处都没有的物种，与美洲大陆上的物种会如此相似？它们的生存条件、地质特征、海拔或是气候，或

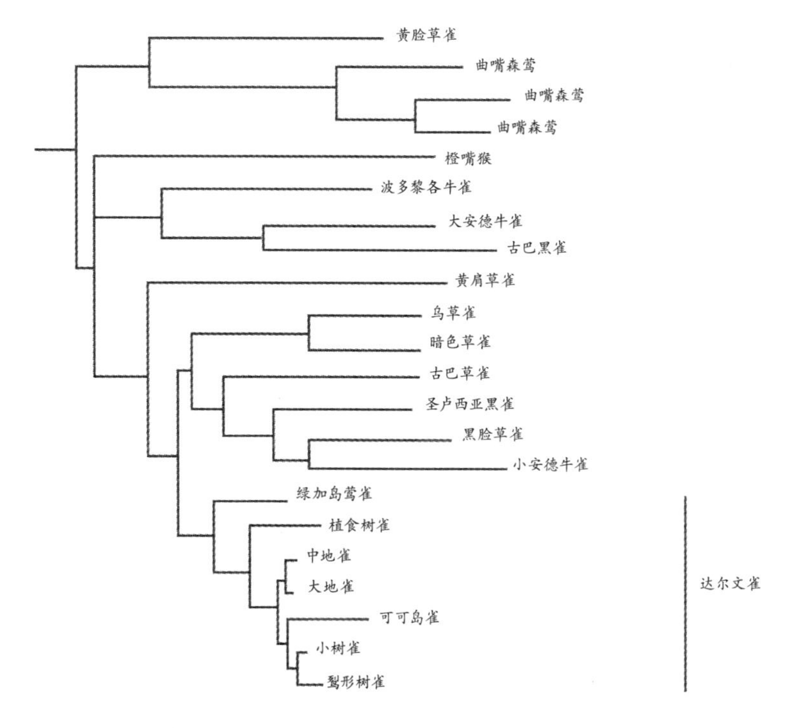


图13 达尔文雀类与它们近亲的系统发育树。这棵发育树是基于不同物种线粒体中一段基因的DNA序列的差异。水平分支的长度说明不同物种间差异的大小（从最接近物种的0.2%到最疏远物种的16.5%）。系统发育树表明加拉帕戈斯群岛上的物种很显然有一个共同的祖先，它们都有着相似序列的这个基因，与距今很近的祖先的序列相一致。与之相对的是，其他具有亲缘关系的雀类物种彼此间的差异要大得多。

是物种的组成结构，与南美洲沿岸的情况都不密切相似。事实上它们之间在所有这些方面都存在相当大的区别。

毫无疑问，进化论为这些问题提供了解释，在过去的150年里对海岛生物的研究已经充分证明了达尔文的高瞻远瞩。

第五章

# 适应与自然选择

## 适应的问题

进化论的一个重要任务就是在生物间不同层次的相似性下解释生物多样性。在第三章中，我们强调了不同类群间的相似性，以及这些现象如何符合达尔文的后代渐变理论。进化论另一个主要的内容就是为生物的“适应”提供科学的解释：它们良好的工艺设计外观，它们与不同生活方式相适应的多样性。这些都使得本章成为本书中最长的一章。

适应的经典例子数不胜数，我们将举出其中的几个来说明问题的本质。不同类型的眼睛的多样性本身就非常令人惊讶，不过与不同动物所生活的不同环境相联系来看就可以理解。在水底使用的眼睛与在空气中使用的不同，捕食者的眼睛具有特殊的适应性，能够看穿被捕食者进化出的伪装。许多水底的捕食者捕食透明的海洋生物，它们的眼睛有着特殊的增强对比度的功能，包括紫外线透视以及偏振光透视。另一个著名的适应例子是鸟类翅膀中中空的骨头，它们的内部支杆与飞机机翼非常相似（图14）；还有就是动物关节处精巧的结构，其表面使得移动的部分能够彼此顺滑地移动。

动物与食性相关的适应以及它们所捕食猎物的反馈性适应

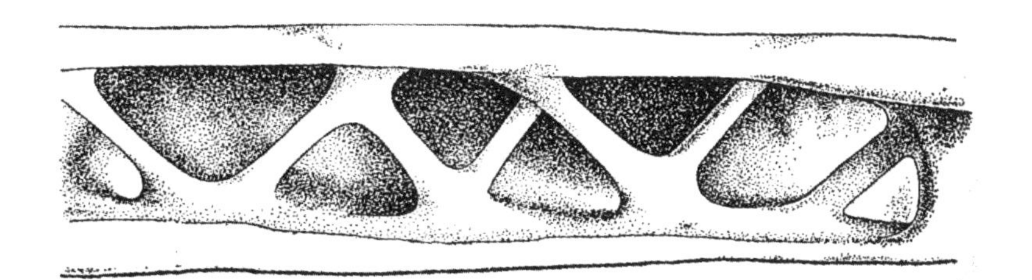

图14 秃鹫中空的骨骼，以及它内部加固的支杆。

还有许多其他例子。蝴蝶拥有长长的口器，用来直达花朵的深处、吸食花蜜；相应地，花朵用艳丽的颜色与特殊的气味吸引昆虫，并为它们提供花蜜。青蛙与变色龙有着长长的舌头，能够通过黏性的舌尖捕食昆虫。许多动物的适应性能够帮助它们逃避捕食者，其外表则取决于所生存的环境。许多鱼类拥有银色的外观，这使得它们在水里不容易被发现，但是陆地动物很少会有这样的颜色。一些动物有着隐蔽色，与树叶或树枝，或者其他有毒有刺的动物颜色相近。

适应性在动物、植物以及微生物的许多细节部分同样能发现，包括每一个层次，小到细胞的机制及其控制（我们在第三章中讲述过）。例如，细胞分裂与细胞迁移是由蛋白分子组成的微小发动机所驱动。遗传物质在产生新的细胞时被复制，此时新产生的DNA会进行校对工作，这大大降低了有害突变发生的概率。细胞表面的蛋白复合物选择性地允许某些化学物质透过，而阻止了另一些化学物质的进入。在神经细胞中，这种控制被用于调节穿越细胞表面的带电金属离子流，从而产生沿着神经传递信息的电信号。动物行为模式是它们神经活动模式的最终输出结果，无疑是对它们生活方式的适应。例如，在鸟类中，巢寄生性鸟类例如杜鹃会将宿主原本的蛋或幼鸟移出巢穴，使得宿主抚养它们的后代。相应地，宿主鸟类变得更为警觉以适应这一现象。种植真菌“花园”的蚂蚁进化出了一些行为，包括清除掉污染它们腐烂叶子的真菌孢子。甚至生物的老化速率都与动植物生长环境相适应，这一点我们将在第七章中进行阐释。

在达尔文与华莱士之前，这种适应性似乎是由造物主创造的。似乎没有其他方法能够解释生物体各方面令人惊讶的精细而完美的细节，正如一块手表的复杂程度不可能是纯自然的产物。18世纪神学家们提出“创世论”来“证明”造物主存在，其主要依据就在于没有其他解释，而“适应”一词的提出是用来描述生物都拥有对它们有用的结构这一现象的。我们要明白，将这一现象表述为“适应”将导致一个问题。认识到适应需要一个解释，这对于我们了解生命有着重要的作用。



毫无疑问，动植物与其他自然产物如岩石或矿物是不同的，我们在“动物、植物、矿物”[①]这一游戏中已经了解了。但是创世论忽视了这种可能性：在产生矿物、岩石、山川的作用之外，可能还会有自然的过程，它们能够将生物解释为复杂的自然产物，而不需要造物主的参与。对适应来源的生物学解释取代了造物主的观点，并成为后达尔文进化学说的中心思想。在本章中，我们将描述适应的现代理论以及它的生物学原因与基础。这些都基于我们在第二章中所概述的自然选择理论。

---

① “动物、植物、矿物”是维多利亚时代英国的传统文字游戏。一位玩家在心中想象一个物品，另一位玩家要在20个问题后猜出这是什么。但是对于20个问题，第一位玩家只能用“是”或“否”来回答。如果回答是类似于“不知道”这样的答案，那么这个问题将不被计入问题总数内。

## 人工选择与可遗传变异

达尔文最早提出并强调的一个高度相关的观察结果是，人类可以有规律地对生物进行改变，能够产生与我们在自然界所见相同的外表。这通常来源于对具有所需要特征的动植物进行人工选择，或是选择性育种。在相对于进化的化石记录而言较短的时间内即可培育相当大的变化。例如，我们已经产生出各种不同品种的卷心菜，包括一些奇特的品种例如花椰菜或西兰花，它们都是产生巨大花朵、形成巨型头部的突变体，而像球芽甘蓝这样的品种，则具有不同寻常的叶子发育（图15A）。与之相似，许多种类的狗是由人类培育的（图15B），正如达尔文所指出的，它们之间的差异与自然界中两种不同物种间的差异很类似。然而，尽管所有的犬属动物（包括土狼与豺狼）都是近亲，也可以进行杂交，不同品种的狗并不是由不同的野狗物种驯化而来，而是在过去一两千年（几百个狗世代）的人工选择下，来自一个共同的祖先——狼。狗基因的DNA序列基本是狼的序列的子集，而土狼（据化石判断，它们的祖先在100万年前与狼的祖先分离）无论与狼还是狗的差异都比差异最大的狼/狗还要大上两倍。狗与狗之间相同基因的序列差异可能在狗与狼分离开之后产生，这种差异可以用来推断它们的分离的发生时间（见第三章）。结论是狗在远超过1.4万年之前就与狼分离——这个时间由考古证据所证实，但是不超过13.5万年前。

人工选择的成功可能是由于在种群与物种中存在可遗传的变异（我们在第三章中所描述的正常个体间细微的区别）。即使没有任何遗传的概念，人们已经让那些具有他们所喜爱或有用特征的动物进行繁殖，在经过足够多的世代之后这个过程已经

A

B

图15 A. 一些卷心菜栽培品种的变异种。B. 不同品种的两只狗的大小与外形差异。

产生彼此间差异巨大的株系，而它们与最初驯化的祖先形态也截然不同。这清晰地说明驯化物种中的个体彼此间必定是不同的，有许多不同能够传递给它们的后代，这意味着它们是可遗传的。如果这些不同只是因为动物或植物被对待的方式不同，选择性育种与人工选择对于下一代则没有影响。除非这些不同能被遗传，否则只有通过改进培育方式才能提高品种质量。

每个你所能想到的性状都能够在遗传上发生变化。众所周知，犬类的不同品种差异不只体现在外观与大小，还体现在心理特质例如性格与气质上：有一些比较友善，而其他的一些则很凶猛，适合作为看门犬。它们对于气味的兴趣不同，它们有些倾向于叼东西，有些喜欢游泳；智力上也同样存在差异。它们所易感染的疾病也不同，一个著名的例子就是斑点狗容易患上痛风。它们的衰老速率甚至也存在差异，有些品种例如吉娃娃，有着令人惊异的长寿（寿命几乎与猫一样长），而其他的一些品种例如大丹狗，寿命则只有吉娃娃的一半。当然，所有这些特质都会受到环境因素的影响，例如良好的照顾与治疗，但它们依然受到遗传的强烈影响。

相似的遗传差异在其他许多家养品种中也同样出现。另一个例子——不同品种苹果的品质就是可遗传的差异。它们包括了对不同人类需求例如早熟或晚熟、适合做菜或是生食的适应，以及对不同国家的不同气候的适应。与犬类的例子类似，在人工选择进行的同时，其他选择过程也同样在苹果中进行，不是所有令人喜爱的特征都会臻于完美。例如，考克斯苹果是一种非常好吃的苹果，但是它非常容易受到病菌的侵害。



## 可遗传变异的种类

人工选择的成功有力地证明了动植物的许多性状差异是可遗传的。众多遗传学研究证明了在自然界中许多生物同样具有可遗传的性状多样性，包括动物、植物、真菌、细菌及病毒的诸多物种。多样性来源于基因DNA序列的随机突变过程，此过程已得到充分认识，与那些引起家族性遗传疾病的过程（第三章）相类似。这些突变中的大多数可能是有害的，例如人类或家畜的遗传疾病，但是有时候也存在着有益的突变。这些突变已使得动物对于疾病具有抵抗力（例如家兔中兔黏液瘤病抗性的进化）。这些也带来了当今社会的一个重要问题：害虫们进化出了对化学药剂的抗性（包括老鼠对于杀鼠灵的抗性，寄生在人类与家畜身上的寄生虫对于驱虫药的抗性，蚊子对于杀虫剂的抗性，以及细菌中的抗生素抗性）。正是由于它们与人类或动物的福祉息息相关，人们对它们中的许多实例已经进行了非常仔细的研究。

可遗传的差异在人类中也有众所周知的事例。变异可能会表现为“离散的”性状差异，例如我们前文中提到的眼睛与头发的颜色。这些是由单个基因中的差别控制的变异，不受环境因素的影响（或是影响极其微小，例如金发人群的头发被日光所漂白）。诸如此类的常见多样性被称作**多态性**。有些情况例如色盲也是简单基因差别，但是在人群中属于非常罕见的变异。甚至连行为方式都可能会遗传。一个火蚂蚁群应该有一个还是多个蚁后，这可能是由单个基因上的一个差异控制的，这个基因编码的蛋白质连接的化学物质参与个体识别。

“连续的”变异同样在种群的许多特征中十分明显，例如人们的身高与体重的渐变。这种变异通常受环境影响较大。20世纪许多不同的国家中都出现了后代身高的增加，这并不是因为遗传的改变而是生活环境的变化，包括更好的营养条件和童年时期严重疾病的减少。然而，在人类种群的此类特征中同样存在某种程度的遗传因素。这是通过对同卵双胞胎和异卵双胞胎的研究获得的。异卵双胞胎是普通的兄弟姐妹关系，只不过是碰巧在同一时间怀胎，他们之间的差异与任何兄弟姐妹一样；但是同卵双胞胎来自于一个一分为二的受精卵，在遗传学角度上是完全一样的。人们业已证明，同卵双胞胎比异卵双胞胎在许多特征上都更为相似，这肯定是由于他们的遗传相似性（当然，要注意对同卵双胞胎的照顾方式不可以比异卵双胞胎的更相像；例如，应该只研究相同性别的两种双胞胎）。尽管环境影响非常重要并且明显地经常存在，各式各样的证据都一致表明许多特征变异都需要一定程度的遗传基础，包括智力方面。人们已经在众多生物的各类特征中验证了可遗传变异的存在。甚至连动物在阶级中的位置，或者说社会等级，也是可遗传的；这种现象已经在鸡群与蟑螂中得到体现。连续遗传变异性的大小可以通过不同程度的近亲间的相似性进行测定。这在动物与农作物育种中发挥了很大作用，饲养员们能够通过这种方法预测不同亲本产生后代的性状，例如奶牛的产奶量，由此对育种进行规划。

遗传差异归根到底就是DNA"字母"的差异。这通常不会导致蛋白质的氨基酸序列改变。当不同个体间相同基因的DNA序列进行比较时，我们就能够发现差异，尽管与不同物种间的序列差异（第三章中讨论过这种差异，见图8）相比通常要小得多。例如，第三章中提到的，可以将来自不同个体的葡萄糖—6—磷酸脱氢酶的基因序列进行比较，可能不存在任何差异（那么就没有多样性）。如果有些种群中的个体存在变异的基因序列，那么在部分比较中将会展现出差异。这被称作分子多态性。遗传学家通过比较种群中个体间存在差异的DNA序列的小片段，对这种多样性进行测定。在人类中，当比较不同人之间相同的基因序列时，通常我们会发现不足0.1%的DNA字母存在不同，而与之形成对比的是人类与黑猩猩间的差异通常达到1%左右。在一些基因中多样性较高，而在另一些中则相对较低，正如我们所预测的，那些可能不那么重要的区域的不编码蛋白质的基因变异通常要高于编码蛋白质基因的变异。与大多数其他物种相比，人类的变异性相当之匮乏。例如，DNA多样性在玉米中更为常见（超过2%的玉米DNA字母是可变的）。

物种中变异性的分布能够为我们提供有用的信息。要繁殖不同性状的狗时，只有亲本的性状十分一致，才能进行育种。这是由于严格的纯血统规则，它对犬类的杂交进行控制，禁止不同品种间出现“基因流动”。一个品种所需要的特性，例如衔回猎获物，便只会在这一个品种中进行充分培育，不同的品种彼此相异。这种品种间的隔离是不符合自然规律的，不同品种的狗能够愉快地进行交配并产下健康的后代。狗的许多变异性相应地是在品种间产生的。许多自然界的物种生活在不同的地理隔绝的种群中，正如我们所预料的那样，此类物种作为整体的多样性比生活在一起的单一种群中的多样性要大得多，因为在种群间存在着差异。例如，某些血型在某些人种中更为常见（见第六章），对于许多其他基因变体来说也是如此。然而，与犬类品种不同，在人类及自然界的其他许多物种中，种群间差异与种群内部的多样性相比要小得多。这种差异是因为人类在种群间能够自由移动。这些基因结果的一个重要含意就是，人类种族是由我们基因组中的极小部分基因区分开的，在全球范围内，我们大部分的遗传结构的变异有着相似的范围与异质性。现代社会愈来愈高的移动性正在快速降低种群间的任何差异。



## 自然选择与适应

自然条件下进化论的一个根本理论就是，一些可遗传的性状差异影响着生存与繁殖。例如，正如为了速度而对赛马进行筛选（通过将冠军马与其近亲进行杂交），羚羊也天生就被用速度筛选过，因为只有那些不被捕食者吃掉的个体才能繁殖下一代。达尔文与华莱士意识到了这个过程能够解释对自然条件的适应。我们通过人工选择改造动植物的能力取决于这种性状是否可遗传。如果存在可遗传的差别，那么自然界中那些成功的个体同样也会把它们的基因（通常还有它们优秀的特质）传递给下一代，而下一代就相应具备了适应性的特质，例如速度。

为了简洁，也为了能让人用普通名词思考，“适应性（fitness）”一词经常会被用于生物学写作中，代表生存及繁殖的总能力，不需要详述所提及的是哪些性状（正如我们用“智力”一词来代表一系列不同的能力）。适应性包括生物体的诸多不同方面，例如，速度只是影响羚羊适应性的一个因素。警惕性与发现捕食者的能力也很重要。然而，仅生存是不够的，繁育后代的能力，例如为后代提供保障与照顾，对于动物的适应性而言同样重要；对于开花植物的适应性而言，吸引传粉者的能力尤为关键。因此适应性一词可以被用来描述对范围极广的不同性状的选择。正如“智力”所遭遇的，“适应性”一词的笼统性也使得它引起误读与争议。

为了了解哪些性状可能在生物体的适应过程中发挥重要作用，我们必须深入了解它的生活规律与生存环境。同样一种特性，在一个物种身上可能会使其具有良好的适应性，而对另一个物种则不尽然。例如，对于一只通过隐蔽色来躲避捕猎者的蜥蜴而言，速度并不是适应的重要因素。对于一只居住在树上的蜥蜴而言，善于抓住树枝比跑得飞快要重要得多，因此短腿相比长腿而言，更具有适应性。对于羚羊而言速度是适应性的，但站住不动避免被捕食者发觉也是许多动物躲避猎杀的一种选择。另外一些动物通过吓跑对方来躲避捕食者：例如，有些蝴蝶的翅膀上有眼状班纹，它们能够突然展开而把鸟类吓跑。植物显然不能移动，但它们也有各式各样的方法来躲避被吃掉的命运，例如有些吃起来苦涩，而有些则长满了刺。所有这些不同的性状都可能会提高生物的生存和/或繁殖的概率，从而提升它们的适应能力。

正如我们在第二章中所表明的，考虑到众多性状的遗传变异性，以及环境因素的不同，自然选择不可避免地会发生，而种群与物种的遗传组成将会随时间变化。这种改变通常会以年为单位缓慢进行，因为种群中的一个遗传变体从稀有变成普遍，通常需要许多代的时间。在动植物的繁殖过程中，常会发生严苛的选择（例如，当疾病使某畜群或作物中的大部分病死时），但是改变依然要花费许多年时间。据估计，玉米是在约一万年前被驯化的，但是现代的巨型玉米棒却是近代的产物。尽管进化的改变以年为单位来看非常缓慢，在化石记录的时间尺度上，自然选择造成的改变却是迅速的。有益的性状在种群中传播开来的速率一开始可能极低，用时则短于地层中两个相邻层之间的时间（一般至少几千年，见第四章）。

尽管相对于我们的生命而言，自然选择发生得太过缓慢，我们通常看不到它的发生，但是自然选择从未停止。甚至我们人类也还在进化。例如，我们的饮食结构已与我们的祖先不同，因此尽管我们的牙齿并不十分坚硬，但是它很适应现代柔软的食物。许多现代食物的高含糖量容易导致蛀牙，甚至是致命的脓肿，但是坚硬的牙齿已不是自然选择所必须的了，因为牙医能够解决这些问题，或者是换上假牙。正如其他如今不再有强烈需求的功能可想而知会发生改变，我们的牙齿可能有一天也会退化。我们的牙齿已经比我们的近亲黑猩猩要小得多，我们还没有阻止它们继续变小的理由。我们的饮食中过量的糖分还导致继发性糖尿病发病率的上升，这种病的致死率非常高。过去，这种疾病主要发生在过了育龄的成人身上，但是现在发病的年龄正持续提前。因此，为了适应我们饮食习惯的改变，一种新的（或许还很强烈）选择压力正趋向于改变我们的代谢特征。在第七章中，我们将展示这些人类生活中的改变是如何使人们进化得越来越长寿的。

适应性的概念经常被人们误解。当生物学家尝试对这个词进行解释时，他们通常会使用与我们日常所说的“适应”相关的例子，例如羚羊的速度。如果我们举鸟类那中空而由支杆交叉强化的轻质骨骼做例子，可能就不那么容易混淆（图14）。自然选择理论是这样解释这种看起来设计精良的结构的：当飞行能力进化时，有着更轻便骨骼的个体的生存几率会比其他个体略微高一些。如果它们的后代继承了这种轻便的骨骼，那么在数代之后的种群中这个特征就会增多。这与人工育种是相似的，在人工育种中饲养员们筛选跑得最快的狗，最终使得所有的灵缇犬有着纤长的腿部。这种腿在跑起来时比短腿更有效率，灵缇犬的腿与羚羊或其他跑得快的动物的腿十分相像，而这些动物都是在自然选择下进化而来的。就算不引入“适应性”的概念，我们也能准确地对自然选择与人工选择进行描述。自然选择就意味着特定的可遗传变体可能会优先被传递给后代。携带有削弱生存或繁殖能力基因的个体，较携带有提升生存或繁殖能力基因的个体而言通常没有那么多机会将基因传给下一代。“适应性”仅仅是一个有用的缩略词，用来概述“性状有时会影响生物的生存和/或繁殖概率”这一思想，且不用特意指出某个性状。在建立自然选择影响种群基因组成的数学模型时，这个概念也非常有用。这些模型的结论为本章的许多论述提供了严谨的基础，不过我们在此不做赘述。

为了说明对于有益突变的选择，我们可以将目光投向人类与老鼠的“军备竞赛”。我们尝试各种针对老鼠的毒药，而老鼠则进化出抗性。杀鼠灵通过阻止凝血来杀死老鼠。它抑制了维生素K代谢过程中一种酶的活性，而维生素K对于凝血与其他许多功能都十分重要。有抗性的老鼠一开始十分稀少，因为它们的维生素K代谢被改变了，这降低了它们生长与存活的概率。换句话说，这就是产生抗性的**代价**。然而，在施用了杀鼠灵的农场与城市中，只有那些有抗性的老鼠能够存活下来，因此尽管需要付出代价，自然选择的力量依然很强大。由此，带有抗性的基因在老鼠种群中传播至很高的携带率，尽管此基因的副作用使得这个基因不能传播到每一个个体。然而，近期的情况是进化出了一种新型的似乎无副作用，甚至可能有益（没有了毒性）的抗性。因此，在老鼠的生存环境改变的情况下，进化将会持续发生。

变异与选择在许多系统中都十分常见，不仅仅是生物个体。遗传物质中某些特定的组分被保留下来，并不是由于它们能够增强携带它们的生物的适应性，而是由于它们可以在遗传物质本身当中复制增殖，就像生物体中的寄生虫。人体中有50%的DNA被视为归于此类。另一个人体中自然选择驱动进化改变的重要例子是癌症。癌症是一种细胞无视身体其余部分利益、自顾自无限增殖的疾病。这种疾病通常由一种能够增大其他基因突变概率的突变（例如，第三章中所提到的校对系统失效，这种系统检查DNA顺序，阻止突变）所导致。一旦突变发生的频率增大，其中的一些将影响细胞的增殖速率，则可能会出现一个快速增殖的细胞系。随着时间的推移，携带有其他基因突变的细胞不断增殖出越来越多的细胞，生长越来越快，最终癌症通常变得越来越严重。癌症细胞同时还能对抑制它们生长的药物产生抗性。如同众所周知的艾滋病患者中艾滋病病毒的抗药性进化一样，获得突变从而免受药物抑制的癌症细胞同样能比初始类型的细胞长得更快，进而使癌症无法缓和。这就是为什么在病情缓解后重新开始药物治疗效果往往甚微的原因。

在另一个极端，拥有不同性状的物种的灭绝速率可能存在不同，即在物种层面上可能也存在着选择。例如，个头较大的物种的种群规模与繁殖速率通常较低，相对于个头较小的物种而言更容易灭绝（见第四章）。与之相对，相同物种中，不同个体间的自然选择通常更青睐较大的个头，这可能是由于较大的个体在食物或配偶的竞争中占据更大的优势。相关物种身体大小的一系列分布情况可能是两种不同类型的选择共同作用的结果。然而，物种内部对个体的选择也许是最为重要的因素，因为它首先产生了不同大小的身体，而且它发挥作用通常比物种层面的选择要快得多。

选择对于非生物事件而言也十分重要。在设计机器与电脑程序时，想要达到最优设计，一种非常有效的方法就是不断地对设计进行随机而微小的调整，保留下效果良好的部分，删除其他部分。这种方法越来越多地被运用于解决复杂系统的设计难题。在这个过程中，设计师不考虑整体规划，只考虑所需要的功能。

## 适应与进化史

自然选择的进化理论将生物的特性解释为连续变化的积累结果，每种变化都提升了生物的存活率或是繁殖成功率。哪些改变可能发生取决于生物的先前状态：突变只能在一定范围内对动植物的发育进行修饰，这个范围是由形成成熟生物的现有发育程序所限定的。动植物育种人所进行的人工选择的结果说明，改变身体部分的大小与形状，或是明显改变生物的外在特征如外表颜色（例如在狗的不同品种中），相对而言较为容易。显著的改变很容易由突变引发，实验遗传学家们也很容易创造一种老鼠或是果蝇株系使其与正常形态之间的差别比野生种类彼此之间的差别大得多。例如，在实验室中，我们可以制造出一只有着四只翅膀而不是正常的两只翅膀的果蝇。然而，这些重大改变通常会严重影响生物的正常发育，降低它们的生存与繁殖成功率，因此不大可能被自然选择所青睐。甚至连动植物育种人都会避免此类现象的出现（尽管这类突变已经被用于培育不寻常的鸽子与狗，这些动物的健康对于育种者来说没有对于农民那么重要）。

由于上述原因，我们推断进化将向着先前的方向微调前进，而不是突然跳跃到一个全新的状态。这在那些需要许多不同组件共同调整的复杂特征，例如眼睛（我们将在第七章中详细讨论）中体现得尤其明显。如果其中一个组件发生了彻底的改变，即使其他部分未发生改变，它们的协作也将受到影响。新的适应性进化时，通常都是在原先结构上的修改版本，而且一般最开始并不会处于最佳状态。自然选择就如同一个工程师，对机器修补、改正以提高性能，而不是坐下来计划好全新的设计。现代的螺丝刀能够用于精密的加工，它有一系列的刀头能够适用于不同的用途，但是螺钉的祖先只是一个由大钉通过一端孔洞旋转的粗纹螺栓。

尽管我们经常惊讶于生物的适应性的精密与高效，它们之中依然存在许多笨拙的修补——一些只有放在它们祖先身上才能理解的特征告诉了我们这一点。画家用肩上的翅膀来表示天使，使得它们能够继续使用上肢。但是所有真实存在的能飞或能滑翔的脊椎动物的翅膀都是改良的前肢，因此翼龙、鸟类以及蝙蝠，都不能够使用前肢的大部分原始功能。类似地，哺乳动物心脏与循环系统有着神奇的特征，反映着这个系统从起源至今逐步修补的历史。最初在鱼体内从心脏泵出血液到达鱼鳃，然后再到达全身（图16）。循环系统的胚胎发育清晰地透露了它进化层面的祖先。

有些时候，在不同的类群中，针对同一个功能性的问题，可能会独立进化出相似的解决方案，导致十分相似的适应性，然而由于不同的进化历史，它们的细部特征又大为不同，例如鸟类与蝙蝠的翅膀。因此，尽管不同生物具有相似性通常是由于它们



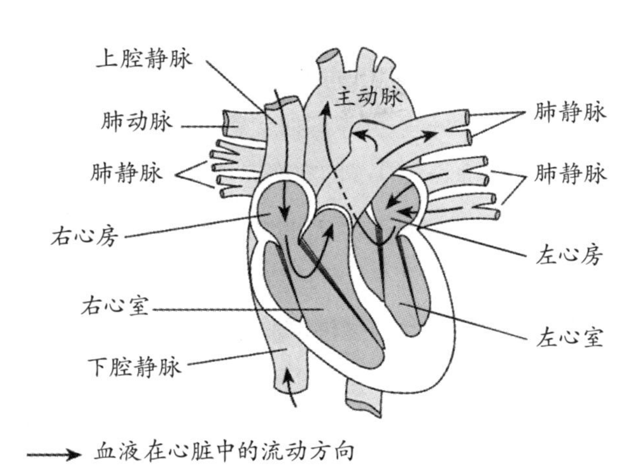


图16 哺乳动物心脏与血管高度复杂的结构。注意肺动脉（输送血液至肺部）笨拙地扭曲在主动脉（将血液输送至身体其他部位）与上腔静脉（将脑部的血液运送回心脏）之后。

具有亲缘关系（如同我们与猿），两个亲缘关系很远的物种生活在相似的环境中有时也会比亲缘关系更近的物种看起来更为相像。如果被这种形态学上的相似与差异误导，可以通过DNA序列的相似与差异发现它们真实的进化关系，正如我们在第三章所述。例如，几种不同的江豚在世界上几个不同区域的大河里进化生存。它们共有某些与公海中物种区别开来的特征，特别是简化的眼睛，因为它们生活在混浊的水中，更多地依靠回声定位而不是视力导航。DNA序列比对结果表明，某种江豚物种与和它生活在同一个区域的海豚间的关系，比它与生活在其他地方的江豚间关系更为紧密。相似的环境导致相似的适应是说得通的。

尽管存在许多相似性，自然选择与人工设计过程依然存在几点差异。一点就是进化是没有前瞻性的；生物只针对一时的主要环境情况发生进化，这样产生的性状也许会在环境剧烈改变时导致它们的灭绝。正如我们将在本章稍后部分所展示的，雄性之间的性竞争将产生一些严重减弱它们生存能力的结构；很有可能，在某些情况下，环境变得不利于生存，存活率降低至这个物种最终无法继续维持下去，拥有多只鹿角的爱尔兰大角鹿就是这样灭绝的。长寿生物的生育力通常进化至极低水平，例如秃鹫这样的猛禽，每两年产下一个后代（我们将在第七章中深入讨论）。如果环境适宜的话，这些种群将会生活得很好，繁殖母禽的年死亡率也很低。然而，一旦环境恶化、死亡率升高，例如遭到人类侵扰，这可能会导致种群数量的急剧减少。现在这种情况依然发生在许多物种身上，已经导致了许多曾经数量巨大的物种的灭绝。例如，在19世纪，繁殖缓慢的北美旅鸽因捕猎灭绝，尽管它最初的数量曾达到几千万只。有些物种进化成为极为特殊的栖息地的领主，但是一旦由于气候原因，这块栖息地消失，它们也同样容易灭绝。例如，中国熊猫的生存受到威胁，因为它们繁殖缓慢，而且以一种只生长在特定山区的竹子为食，而这种竹子现在正遭到砍伐。

自然选择同样并不必然产生完美的适应。首先，可能没有时间将一种生物机制的各个方面调整到最好的状态。当选择的压力来源于一对物种（例如宿主与寄主）间的相对作用时，这种现象将更为明显。例如，宿主抵抗感染能力的加强增大了寄主克服这种抗性的选择压力，强迫宿主进一步进化出新的抗性，如此循环往复。这就是进化的军备竞赛。在这种情况下，没有任何一方能够长时间保持绝对的适应。尽管我们的免疫系统抵抗细菌与病毒侵扰的功能卓越，我们依然容易受到最新进化的流感与感冒病毒菌株的侵害。其次，正如我们之前所提到的，进化的修修补补的特性，即只能在已经产生的东西上进行调整，限制了进化所能达到的效果。脊椎动物眼睛中负责从光敏细胞中传导信号的神经位于视网膜细胞的前部而不是后部，这从设计角度来看，似乎十分可笑，但是这是由于眼睛的这个部分是作为中枢神经系统的分支发育而来，这种发育方式最终造成了这样的结果（章鱼的眼睛与哺乳动物的类似，但是安排要更为合理，它的光敏细胞位于神经的前部）。第三，一个系统某一方面功能的提升可能会造成其他方面功能的减弱，正如我们讨论对杀鼠灵的抗性时提到的。这种情况可能会阻碍适应的改良。我们将在本章后面的内容以及第七章中讨论衰老时提到一些其他的例证。

## 发现自然选择

达尔文与华莱士在不了解自然选择在自然界中产生作用的例证的情况下，提出自然选择是适应性进化的原因。在过去的

50年间，人们发现了许多自然选择的实例并进行了仔细研究，有力地支持了该学说在进化论中的中心地位。我们在此只讨论其中的几个例子。现代社会中一种非常重要的自然选择正在使细菌对于抗生素产生日益增强的抗药性。这是一个被重点研究的进化改变，因为它威胁到了我们的生命，同时发生得非常迅速且（很不幸地）反复出现。在笔者写下本段文字的这天，报纸的头条就是：在爱丁堡皇家医院里发现了具有甲氧西林抗性的葡萄球菌。抗生素一旦被广泛使用，不久就会出现有抗性的细菌。抗生素在1940年代被首次广泛使用，之后不久微生物学家们就提出了对于细菌抗药性的担忧。《美国医学杂志》（其受众主要为医生）上的一篇文章就写道：对于抗生素的滥用“充满了对具有抗性的菌株进行筛选的风险”；在1966年（那时人们还没有改变他们的做法），另一位微生物学家写道：“难道没有办法引起普遍关注，以对抗生素抗性发起反攻吗？”



抗生素抗性的迅速进化并不令人惊讶，因为细菌繁殖非常迅速，且具有庞大的数量，因此任何能够使细胞产生抗性的突变都必定会发生在某个种群的某些细菌中；一旦这些细菌能够在突变带来的细胞功能改变下存活下来并且繁殖，一个具有抗性的种群就会迅速建立起来。人们可能希望抗性对于细菌而言代价昂贵，在老鼠对于杀鼠灵的抗性中最初的确如此，但是对于老鼠，我们不能指望这种情况持续太长时间。细菌迟早会进化得能够很好地适应当前抗生素且自身不付出重大代价。因此，我们只有少量使用抗生素，保证它们只用在确实必要的情况下，并确保所有的感染细菌都在还来不及进化出抗性前就被迅速杀灭。如果在一些细菌还存活的情况下就停止治疗，它们的种群中不可避免地就会包含一些具有抗性的细菌，这些细菌就可能会感染其他人。对抗生素的抗性还可以在细菌间，甚至是不同物种的细菌间传递。对家畜使用的用来减少传染病以及促进生长的抗生素能够引发抗性传播至人类病原细菌。甚至这些后果都不是问题的全部所在。具有抗性突变的细菌不是它们种群之中的典型代表，但是在一些情况下它们有着高于平均水平的突变率，这使得它们能够对选择压力更快地响应。

不论何时，只要人们用药物去杀灭寄生虫或是害虫，对药物或杀虫剂的抗药性就会被进化出来。事实上，人们已经对成百上千例微生物、植物、动物的案例进行了研究。当艾滋病人使用药物进行治疗时，甚至艾滋病毒都会突变，进化出抗性使得治疗最终失效。为了避免这种情况发生，经常使用两种而不是一种药物进行治疗。因为突变是小概率事件，病人体内的病毒种群不大可能同时迅速获得两种抗性突变，但是最终，这种情况通常还是会发生。

这些都是自然选择的实例，但就像人工选择一样，自然选择也包括环境受到人为因素干扰而改变的情况。许多其他人类活动正在引起生物的进化改变。例如，为了象牙猎杀大象的行为似乎已经导致了大象中无牙品种的增多。在过去，这些大象属于罕见、畸形动物。现如今，猖獗的猎杀行为使得这些不寻常的品种能够较正常物种有着更高的生存与繁殖几率，结果导致它们在大象种群中比例的上升。又比如，小翅膀的燕尾蝶飞行能力很差，但是在一些碎片化的栖息地中，或许由于这些飞不远的个体更可能留在适合生存的栖息地里，因此被自然选择青睐。当人们清除花园或是农田里的杂草时，也在对这些一年生植物的生存历史进行选择，使得它们更加迅速地产生种子。对于早熟禾这样的物种，存在着发育更缓慢的个体，它们可以生存两年甚至更久，但是这在密集除草的情况下将成为明显不利的因素。这些例证不仅展示了进化改变有多普遍而迅速，同时也说明我们所做的任何事情都有可能影响与人类有关的物种的进化。鉴于人类遍布于地球，极少有物种能不受到人类的影响。

生物学家同样研究了许多纯粹的不涉及人类栖息地退化或改变的自然选择情况。其中最好的例子之一就是皮特与罗丝玛丽·格兰特在加拉帕戈斯群岛达夫尼岛上有关达尔文雀类中的两个物种（地雀与仙人掌雀）长达30年的研究（见第四章）。这些物种的鸟喙平均尺寸与外形各不相同，但是每个物种的这两种性状都有相当的变异。在研究过程中，格兰特的团队有计划地为岛上每一只鸟戴上环志，并测量它们的鸟喙，对每一只雌鸟的后代也都进行了识别。研究者们跟踪这些后代的幸存情况，并与对它们身体各相应部分的尺寸与外形的测定结合起来。谱系研究表明鸟喙特征的变异与遗传有很大关系，因此后代与亲本相似。对于鸟类野外食性的研究表明，鸟喙的尺寸与形状将影响鸟类处理不同类型种子的效率：大而深的鸟喙能够更好地咬开大的种子，而对于小种子而言，小而浅的鸟喙更适用。受厄尔尼诺现象影响，加拉帕戈斯群岛常有严重干旱现象，而干旱将影响到不同类型食物的数量。在干旱的年份，除了一种种子特别大的物种外，大多数植物都不能产生种子。这意味着有着又大又深的鸟喙的鸟类比其他种类有着大得多的生存机会，这在种群数量统计中有了直接体现：在旱季之后，两个物种中存活下来的成年鸟类比起旱季之前都有更大且更深的鸟喙。此外，它们的后代也遗传了这些特征，因此这个由干旱造成的选择方向上的改变引起了种群组成的遗传性改变——真正的进化改变。考虑到亲本与后代之间的相似程度，这种改变的幅度符合通过观测死亡率与鸟喙特征间的联系所推断的结果。一旦环境恢复到正常状况，鸟喙特点与死亡率间的关系也发生了变化，大而深的鸟喙不再具有优势，而种群数量也后退到了之前的情况。然而，即使在不干旱的年份里，环境中依然存在许多微小的变化，它们将导致鸟喙与适应性间的关系出现变化，因此在整个30年间，鸟喙的特征一直波动，两种鸟类的种群数量最终都与一开始时有显著不同。

花朵对昆虫及其他传粉者的适应是另一个很好的例子。对一株将与同一物种的其他植株进行交配的植物来说，必须吸引传粉者来拜访它们的花朵，并给予这些传粉者奖励（用可食的花蜜或是额外的花粉），以保证它们能够再去拜访同种的其他植物。无论是植物还是传粉者，在此互动中都在进化，为自己争取最大的利益。例如，对于兰科植物，为了让花粉块能够在传粉蛾子来访时牢牢地贴附在它们的头部，让这些蛾子能够深入到花朵内部很重要。这可以使得花粉块在蛾子拜访下一朵花时，准确落到花朵的合适部位，使花朵成功受精。这需要花朵的花蜜差不多恰好处在蛾子的口器所能到达的范围之外，这种需求驱动了对蜜腺管长度的自然选择，于是蜜腺管长度异常的花朵的受精概率将降低。蜜腺管太短的花朵会使得蛾子不用拾起或储存花粉就可以吮吸到花蜜，而蜜腺管太长的花朵将浪费花蜜，就像一盒果汁，它所附的吸管总是太短而不能把盒子中的果汁全部吸出。在果汁盒子行业，这种浪费将会造福果汁销售商，使得他们能够卖出更多果汁，但是对于植物来说制造无用的花蜜将丢失能量、水分与营养，这些资源本应该用在更需要的地方。

一种生活在南非的剑兰每株植物只有一朵花，有着更长蜜腺管的个体比一般个体更容易产生果实，同时每个果实中的种子数量也比一般个体要多。这种植物的蜜腺管长度平均为9.3厘米，而它们的传粉者天蛾的口器长度在3.5—13厘米之间。没有携带花粉的蛾子都拥有最长的口器。这个地区其他不为这种植物传粉的天蛾物种的口器长度平均不足4.5厘米。这说明选择的力量使得花朵与蛾子都去适应彼此，达到某些情况下的极值。有一些生活在马达加斯加的兰花种类的蜜腺管长度甚至达到30厘米，而它们传粉者的口器则长达25厘米。在这些物种中，已经有实验演示对长度的自然选择，在实验中蜜腺花距被打结以缩短其长度，使得蛾子带走花粉块的几率降低。

类似的选择与反选也影响着我们与寄生虫的关系。人们已对若干种人类适应疟疾的方式进行了深入研究，也已经明显进化出许多不同的防御方法，其中就包括在复杂的生命周期的某些阶段，疟原虫生活的红细胞发生的改变。与老鼠产生杀鼠灵抗性的情况类似，这种防御办法有时也会带来一定的副作用。镰刀形红细胞贫血症是一种细胞中血红蛋白（红细胞中主要的蛋白质，作用是在体内携带氧气）改变造成的疾病，如果不医治容易造成死亡。它的变化形式（血红蛋白S）是正常成人血红蛋白A编码基因的一种变体形式，两者之间存在一个DNA字母的差异。为此蛋白编码的一对基因如果都是S型的话，个体将患上镰刀形红细胞贫血症，其红细胞将变得畸形、造成微血管的堵塞。拥有一个正常的A型基因与一个S型基因的人不会感染疾病，而且对疟疾的抵抗能力要高于拥有两个A型基因的人。拥有两个S型基因会造成的疾病就是人们对疟疾的抗性所付出的代价，这使得S型基因不能在人群中传播开来，即使在疟疾高发地区也是如此。同样能够帮助抵御疟疾的葡萄糖—6—磷酸脱氢酶变体（见第三章）也伴随着代价，具有这些变体的人们吃下某些食物或药物，将导致红细胞受到损害，而不具抗性的个体则不会发生这种情况。然而，那些没有代价或代价甚小的疟疾抗性依然是存在的。达菲阴性血型系统是血红蛋白的另一种特征，在非洲的大部分地区广泛分布。相较于达菲阳性个体，拥有达菲阴性血型的人们不易感染特定类型的疟疾。

对于疟疾的抗性说明了一个普遍的认知，即在同一个选择压力下（在上文的例子中是一种严重的疾病），可能会产生不同的响应。有些对疟疾的响应方式比其他方式要好，因为他们对当事人造成的伤害更小。事实上，在不同人类种群中可以发现许多不同的对于疟疾具有抗性的遗传变异，而在某个区域哪些特定类型的突变能够被选择确立下来大体上似乎是一个随机事件。

上文中所讨论的实例说明了自然选择对于人类与动植物生存环境的改变产生的响应。或许出现了一种疾病时，人群中会出现选择，于是进化出有抗性的个体。又或是一只蛾子进化出更长的口器从而能够从花朵中汲取花蜜而不用携带花粉，如此一来花朵反过来也会进化出更长的蜜腺管。在这些例子中，自然选择改变了生物，正如达尔文在1858年提出的设想（见本书第二章所引用的）。然而，自然选择同时经常会阻止改变的发生。在第三章中对细胞中蛋白质与酶的作用机制进行描述时，我们提到突变会发生并会减弱这些功能。即使在一个稳定的环境中，自然选择也在一代代个体中发挥作用，对抗着突变基因（这些基因为突变的蛋白质编码，或是让它们在错误的时间、地点表达，或是表达数量不对）。在每一代中，都会产生具有突变的新个体，但是非突变个体倾向于产生更多后代，因此它们的基因始终最为普遍，而突变个体则在种群中保持较低的水平。这就是

**稳定化**选择或者说**净化**选择，它使得一切尽可能好地运行。在血液凝结中的一类蛋白的编码基因就是其中的一个例子。蛋白序列的某些改变将会导致个体在受伤后无法凝血（血友病）。直到不久之前人们才发现血友病的发病机理，从而能够通过注射凝血因子蛋白帮助血友病患者。在此之前，这种疾病通常会致死或是严重降低生存概率。遗传医学家们已经描述了成千上万种类似的对人体有害的低频基因变异，涵盖了每一种能想象到的性状。

如果环境保持相对稳定，自然选择有足够的时间调整生物性状至能带来高度适应性的状态，那么就发生了稳定化选择。如今，在生物持续变异的性状中我们可以探测到这一选择在发挥作用。人类出生时的体重就是一个例子，相关研究已经十分成熟。即使在新生儿死亡率非常低的今天，中等体重的婴儿的存活率依然是最高的。不高的新生儿死亡率主要涵盖那些太小的婴儿，以及某些太大的婴儿。稳定化选择也发生在动物之中，例如在严重的暴风过后，存活下来的鸟类和昆虫的大小都趋向于中等，最小和最大的往往消失。即使是对最适值的微小偏差也可能会降低生存或繁殖的成功率。因此，生物对于它们所生存的环境的适应能力往往惊人是可以理解的。正如我们在第三章中所提到的，有时候，再微小的细节也可能会发挥重要的作用。生物经常能达到接近完美的状态，例如蝴蝶伪装成树叶或毛毛虫伪装成为树枝这些异常精密的拟态。稳定性选择同样解释了为什么物种往往显示不出进化方面的改变；只要生存环境不存在新的挑战，选择就倾向于让事物保持原有的状态。这样也就能理解有些生物在很长一段进化时间里保持稳定的形态，例如被称作**活化石**的生物们，它们的现代种类与它们远古时期的祖先非常相似。

## 性选择

自然选择是对适应的解释中唯一经过实证检验的。然而，选择也不总是增加总生存率或作为整体的种群的后代数量。当资源有限时，能在竞争中占据优势的特征可能会降低所有个体的生存概率。如果最有竞争力的个体种类在种群中普遍出现，那么整个种群的存活率也许会下降。竞争的这种负面结果不只限于生物学情况。某些侵入式的、低俗的、重复洗脑的广告也是众所周知的例子。

生物竞争中广为人知的例子就是雄性获取配偶的竞争。在很多动物中，并不是所有可繁殖的雄性都能够留下后代，只有那些在与其他雄性的斗争之中和/或在求爱行为中获得胜利的才有机会。有些时候，只有“占据统治地位”的雄性才能获得雌性的青睐。甚至连雄性果蝇在获准交配前都需要向雌性求爱——通过跳舞、唱歌（拍击翅膀获得的声音）以及气味。并不是所有时候都会成功，这并不意外，因为雌性十分挑剔，而且不会与非同类的雄性进行交配。在许多哺乳动物，例如狮子中，存在着交配权力的等级制度；雌性十分挑剔，雄性个体的繁殖成功率是不同的。因此，自然选择会青睐那些让雄性在交配等级中更具优势或是增强它们对雌性的吸引力的性状。雄鹿有着巨大的鹿角，它们用鹿角彼此争斗，有些物种还有其他恐吓手段，例如高声的咆哮。如果这些性状能够遗传（正如我们之前所看到的，这种情况很常见），有着能帮助它们成功交配的性状的雄性，会把它们的基因传递给许多后代，而其他的雄性的后代则会较少。

在这样的**性选择**中，两种性别都会进化出相应的特质，这大概也是许多鸟类拥有鲜艳羽毛的原因。然而，对于许多物种而言，这些特质都集中在雄性身上（图17），说明它们的此类特质并不只是为了自身更好地适应环境。许多此类的雄性特征显然并不能增加生存几率，反而由于其雄性隐性基因携带者的低生存率而常常造成负担。雄孔雀拥有巨大而绚烂的尾羽，但飞行能力很弱，如果尾巴能够小一些的话，也许它们能够更快地从捕食者口下逃脱。对于航空空气动力学研究而言，孔雀显然不是理想的研究对象，不过即使对于燕子而言，它的尾巴也比最适宜飞行的长度要长，但长尾巴的雄燕更受雌燕青睐。即使不那么引人注目的雄性求偶特质也常会带来更大的风险。例如，某些热带的蛙类在以歌唱求爱时，会被蝙蝠探测到而捕获。即使没有这些危险，雄性的求偶行为也将花费大量的精力，而它们本可以把这些精力用在例如觅食等方面，到了交配季节的尾声，这些雄性往往都处于精疲力竭的状态。



达尔文意识到了这一点，他认为求偶方面的选择与其他大部分情形下的选择不同，进而引入了一个特殊的名词“性选择”来强调这一不同。正如我们刚刚讨论过的，雄孔雀的尾巴不可能良好适应，原因有二：一是由于客观原因，这种尾巴对于飞行动物来说看起来不是好的设计；二是由于，**如果**这是好的，雌孔雀就也会有。因此似乎这种选择是在孔雀这种交配竞争异常激烈的物种中，用飞行能力的减弱来换取雄性交配几率的升高。因此，性选择再一次体现出生物学中使用的“适应”一词与日常生活中所提到的适应是有所不同的。一只拖着臃肿尾巴的雄孔雀不“适应”好好飞行或是奔跑（尽管如果它营养不良或是不健康的话也长不出这么大的尾巴），但是在进化生物学的简略表达里，它是高度“适应”的；没有了它的大尾巴，雌孔雀就会与其他雄性交配，它的繁殖几率就下降了。

图17 性选择的结果，达尔文《人类起源与性选择》一书中的插图。图片展示了同一品种天堂鸟的雌鸟与雄鸟，图中可以看出雄鸟羽毛华丽，而雌鸟则缺少装饰。

第六章

# 物种的形成与分化

生物学的常见事实之一就是将生物划分为可辨识的不同物种。随便看看生活于欧洲西北部一座小镇上的鸟类，甚至都可以发现很多种类：知更鸟、乌鸫、画眉、槲鸫、蓝冠山雀、大山雀、鸽子、麻雀、燕雀、八哥，等等。每一个种类都有着与众不同的身型、羽毛颜色、鸣叫声、进食和筑巢的习惯。在北美东部，可以找到一系列不同但又大致相似的鸟类。同一种类的雄鸟和雌鸟成双成对，它们的后代当然也跟它们同属于一个种类。在一个给定的地理位置中，有性繁殖的动植物几乎总是可以容易地被划分为不同的群体（虽然有时细致的观察所找到的物种只存在很轻微的解剖学上的差异）。由于异种之间并不杂交，所以共同生活在同一地点的不同物种保持着区别。多数生物学家认为不能杂交（**生殖隔离**）是划分不同物种的最好标准。对于那些不通过有性繁殖产生后代的生物，比如许多种类的微生物来说，情况就复杂得多。这点我们将在后面再进行探讨。

## 物种间差异的本质

尽管，就像习惯于重力一样，我们习惯于将生物体划分为独立的物种，并认为这是理所应当的，但划分物种并不是显然有必要的。很容易想象一个不存在如此明显差异的世界；以上文提到的鸟为例，有可能存在着生物体具有混合的特征，比如，不同比例混合了知更鸟和画眉的特征；对于给定的一对亲代来说，按不同比例交配可以产生带有不同混合特征的后代。如果没有不同物种间的杂交繁殖障碍，我们现在所看到的生物多样性将不复存在，取而代之的是一些接近于连续的形态。事实上，当出于某种原因，已经分离了的物种间的杂交繁殖障碍打破时，确实会产生出如此高度变异的后代。

因而对于进化论者来说，一个根本问题是要解释物种是如何变得彼此不同的，以及为什么会存在生殖隔离。这是本章的主题。在开始这个问题之前，我们先介绍一下近缘种被阻止杂交的一些途径。有时，主要的障碍是物种间生境或繁殖时间的简单差别。以植物为例，每年都有一段典型的短暂花期，花期没有重叠的植物显然不可能杂交。对于动物来说，不同的繁殖场所可以防止不同物种的个体相互交配。即便是有些生物体在同一时间到达了同一地点，那些只有通过对物种生活史的细致研究才能发现的细微特征，常常可以阻止不同物种的个体互相交配。生物体的一些细微特征只有通过对物种自然史的细致研究才能发现，这些特征往往阻止不同物种的个体成功地彼此交配。例如，由于对方没有恰当的气味和声音，一种生物可能不愿意去向其他物种的生物求偶，或是所展现出的求爱方式相异。交配上的行为学障碍在许多动物中都很明显，植物则通过化学手段辨别来自异种生物的花粉并拒绝它。即使发生了交配，来自异种生物的精子也可能无法成功让雌性的卵子受精。

然而，一些极为近缘的物种间偶尔会发生交配，特别是在没有同种个体可供选择的情况下（如第五章中提到的狗、土狼、豺狼）。但在很多此类情形下，杂交第一代常无法发育；异种个体间试验性交配所产生的杂交体，常会在发育早期死亡，反之，同种个体间交配所产生的大部分后代都可以发育成熟。有些时候，杂交个体能够生存下来，但要比非杂交体的存活概率低很多。即使杂交体能够存活下来，通常也是不育的，不会产生能将基因继续传递下去的后代；骡子（马和驴交配产生的杂交体）就是一个著名的例子。杂交体完全地不能存活或是无法生育显然可以将两物种隔离开来。

## 杂交繁殖障碍的进化

尽管人们已熟悉阻碍杂交繁殖的不同途径，却仍然困惑于这些途径是如何进化的。这是了解物种起源的关键。就像达尔文在《物种起源》第九章中指出的那样，异种交配产生的杂交体不能存活或是无法生育极不可能是自然选择的直接产物；如果一个个体与异种交配产生出无法存活或无法生育的后代，这对它来说没有什么好处。当然，杂交后代不能存活或是无法生育，将有利于避免异种交配，但是对于杂交后代可以很好地生存的情况，就很难看出有任何此类益处了。由此看来，物种因地理或生态分离而彼此隔离之后发生进化改变，异种交配中存在的多数障碍可能是这种进化改变的副产物。

比如说，想象一种生活在加拉帕戈斯群岛一个小岛上的达尔文雀。假设少量的个体，成功地飞到了先前未被该物种占领过的另一个小岛，并成功地在此栖息繁衍。如果这样的迁徙是十分稀有的，那么这些新种群和原始种群将会彼此独立进化。通过突变、自然选择和遗传漂变，两种群间的基因构成将会分化。这些变化将由种群所处的并开始适应的环境差异来推进。例如，不同岛上可供吃种子的鸟类食用的植物有所不同，甚至由于岛上食物丰富程度的差异，不同岛上的同一种雀鸟喙的尺寸也存在差别。

一个物种的种群会因所处的地理位置而变化，这种变化通常使其适应所处环境——这一趋向叫作**地理变异**。有关人类的显而易见的例子，是在不同种族间大量存在的微小的体质差异，以及更小的局部特征差别，如皮肤颜色和身高。这种变异性在其他许多生活于广阔的地域环境里的动植物中也可以找到。在一个由一系列地域性种群组成的物种中，一些个体通常会在不同地点之间迁移。但不同生物发生迁移的数量差别极大；蜗牛的迁移率非常低，而一些生物体，如鸟或许多会飞的昆虫，则有着较高的迁移率。如果迁移的个体能够与到达地的种群杂交繁殖，就能为当地种群贡献它们的基因组分。因而迁移是一种均质化的力量，与自然选择或遗传漂变引起地域种群产生基因分化的趋势正相反（见第二章）。一个物种的多个种群间或多或少都会发生分化，这取决于迁移的数量及促使地域种群间产生差别的进化动力。强的自然选择可以导致即便是相邻的种群也变得不同。例如，铅矿和铜矿的开采会造成土壤受金属污染，这对多数植物都有极大的毒害作用，但是，在许多矿山周围受污染的土地上都进化出了对金属耐受的植物。如果没有金属，耐受植物则生长不良。因而，耐受植物只在矿山上或靠近矿山的地方生长，而生长在远离矿山的边缘地带的植物则剧变为非耐受植物。

在那些不那么极端的情况下，物种特征表现出地理性的渐变，这是因为迁移模糊了自然选择所引起的随地理环境变化的差异。许多生活在北半球温带地区的哺乳动物有着较大的体形。动物的平均体形大小从南至北呈现出或多或少的连续性变化，这很可能是在寒冷的气候下，动物为减少热量损失而选择较小的表面积和体积之比的反映。出于类似的原因，相较于南方种群，北方种群还倾向于拥有更短的耳朵和四肢。

不同类型的自然选择，并不是同一个物种各地理隔离种群间出现差异的必要因素。同一种自然选择有时会导致有差别的响应。比如，像第五章中所描述的那样，遭受疟疾感染的地区的人口有着不同的抗疟疾基因突变。有多种分子途径可引发抗性。引发抗性的不同突变会在不同地区偶然出现，抗性突变在特定人群中渐渐占据优势很大程度上是运气使然。即便完全没有自然选择，由于前文中提到的遗传漂变的随机过程，一个物种的不同种群间也会逐步形成差异。许多物种中，不同种群间常常存在着显著的基因差别，即使是在DNA和蛋白序列的变异没有影响可见特征的情况下。在这一点上，人类也不例外。即便在英国，人群中A，B，O血型的出现频率也有所不同，这取决于单个基因的变异形式。例如，相较于英格兰南部，O型血人群在威尔士北部和苏格兰更为常见。不同血型的出现频率在更广泛的区域有着更大的差别。B型血人群在印度部分地区的出现频率超过了30%，而在美洲原住民中却极为罕见。

这种地理变异的例子还有很多。尽管主要的种族间存在着可见的差异，但不同人群或种族群体间并没有生物上的异种繁殖障碍。然而，对于一些物种来说，在最极端的情况下，同一物种不同种群间的差别大到可能会被认作不同的物种，只不过这两个极端的种群由一系列彼此杂交繁殖的中间种群相联系。甚至还存在这样的情况：一个物种的极端种群间的差异大到它们之间无法杂交繁殖；如果中间种群灭绝，它们将会构成不同的物种。

这就解释了一个重要观点：根据进化论，在生殖隔离形成的过程中，必然存在着过渡阶段，因而我们应该至少能够观察到一些难以将两个相关种群进行分类的情况。尽管这给我们将生物按既有方式分类造成了麻烦，但是这是可以预见的进化结果，而且在自然界中也是易见的。两个地理隔离种群，在产生生殖隔离的进化过程中存在过渡阶段的例子有很多。美洲拟暗果蝇是已被人们充分研究的一个例子。这类生活在北美洲和中美洲西海岸的生物，或多或少地连续分布于从加拿大到危地马拉一带，但在哥伦比亚波哥大地区却存在着一个隔离的种群。波哥大果蝇种群看上去跟其他果蝇种群一模一样，但是它们的DNA序列却有轻微的差别。由于序列的差异需要长时间的积累，波哥大种群可能是在约20万年前由一群迁移到那里的果蝇形成的。在实验室，波哥大种群已经可以跟来自其他种群的拟暗果蝇相交配，产生的第一代杂合体雌性是可育的，而以非波哥大种群的雌性作为母代杂交所产生的雄性杂合体却是不育的。非杂合体雄性不育的现象在其他有着较大差异的果蝇种群的交配中从未出现过。如果将果蝇主要种群引进波哥大，它们之间想必会自由杂交繁殖，又由于雌性杂合体是可育的，那么杂交就可以一代代持续下去。因此波哥大种群的与众不同完全是由于地理隔离造成的。因此，尽管雄性杂合体不育显现出波哥大种已经开始形成生殖隔离，但并没有令人信服的理由将波哥大种群单独视为一个物种。

不难理解，就像加拉帕戈斯雀一样，为什么同一物种不同地区的种群，在不同的生活环境下会产生适应各自所处环境的不同特征。但为什么这样会造成杂交繁殖障碍却不太容易理解。有时这可能是适应不同环境而产生的相当直接的副产品。例如，两种生长在美国西南部山区的猴面花植物，彩艳龙头和红龙

**两种沟酸浆属植物花的特征**

| 物种名 | 彩艳龙头 | 红龙头 |
|---|---|---|
| 传粉者 | 蜜蜂 | 蜂鸟 |
| 花的大小 | 小 | 大 |
| 花型 | 宽，有平底 | 窄，管状 |
| 花色 | 粉红 | 红 |
| 花蜜 | 中等水平，高糖 | 丰富，低糖 |

头。和大多数猴面花一样，彩艳龙头由蜜蜂授粉，它的花有着适应蜜蜂授粉的特质（见上表）。与众不同的是，红龙头由蜂鸟授粉，它的花有着利于蜂鸟授粉的几处不同特征。红龙头可能是由跟彩艳龙头外观相近的由蜜蜂授粉的植物，通过改变花的特征进化而来。

两种猴面花可以实验性交配，杂合体健康可育，然而在自然界中两种植物并肩生长却没有杂交混合。野外观察结果显示，蜜蜂在采集过彩艳龙头后，极少会再去采集红龙头，而蜂鸟在采集过红龙头后，极少会再去采集彩艳龙头。为了查明传粉者对有着两种花的特质的植物会如何反应，人工培育了拥有两亲本广泛的混合特征的第二代杂交种群，并将其种植在野外环境中。最能促进隔离形成的特征是花色，红色可以阻挡蜜蜂而吸引蜂鸟的授粉。其他的特征可以影响两授粉者的其中一个。花蜜含量更高的花朵吸引蜂鸟的授粉，而花瓣较大的花朵对蜜蜂的吸引力更大。混有两种特征的中间型既可能被蜜蜂授粉，也有可能被蜂鸟授粉，因而与亲本物种之间产生了中等程度的隔离。在这个例子中，随着蜂鸟授粉进化由自然选择驱动的改变已经使红龙头和近缘种彩艳龙头间产生生殖隔离。

虽然在大多数情况下我们不知道究竟是什么力量促使近缘

种的分化，并最终导致了生殖隔离，但是，如果两个地理隔离种群间存在独立的进化差异，那么两者之间生殖隔离的根源就并不特别令人惊讶。种群基因构成的每一个变化，必定要么是自然选择的结果，要么是能轻微影响适应性并能通过遗传漂变扩散出去（在第二章及本章末尾处有讨论）。如果变异体因为有更强的使种群适应当地环境的能力而在种群中蔓延开来，当它与未曾自然接触过的、来自其他种群的基因相结合（通过杂交）时，这种蔓延不会被任何不良影响所阻拦。任何一种自然选择都不能将地理或生态隔离种群个体间的交配行为的兼容性维持下去，或者使开始在不同种群间分化的基因保持自然进化的和谐关系。就像其他没有因自然选择而维持下去的特征一样（比如洞栖性动物的眼睛），杂交繁殖的能力也会随时间而退化。

如果进化分化程度足够大，完全的生殖隔离看起来是不可避免的。这并不比英国产的电插头与欧洲大陆的插座不匹配的事实更让人惊讶，即便每一种插头都与相应的插座良好匹配。人们必须持续不断地努力以确保设计的机器有良好的兼容性，比如为个人计算机和苹果电脑设计的软件。种间杂交的遗传分析显示，不同物种的确携带有一些不同的基因系列，当这些基因在杂合体中混合后，会出现机能失调的状况。就像上文中提到的那样，许多异种动物杂交后产生的第一代雄性杂合体不育，但是雌性可育。因而可育的雌性杂合体是有可能和两个亲本物种中的某一个杂交的。通过对此类杂交产生的雄性后代的生育力进行测试，我们可以研究雄性杂合体不育的遗传基础。人们已在果蝇物种身上做了大量这类研究；结果清晰地表明，两物种不同基因的相互作用是导致杂合不育的原因。例如，在拟暗果蝇的大陆种群和波哥大种群间有差别的基因中，大约15个在两个种群间有差别的基因看来参与导致了雄性杂合体不育。

两个种群间产生足够造成生殖隔离的差异所需要的时间差别很大。拟暗果蝇用了20万年（超过100万代）的时间，仅仅产生了很不完全的隔离。在其他例子里，有证据表明，生活在维多利亚湖的慈鲷科鱼类有很快的生殖隔离进化速度。虽然地质证据显示维多利亚湖形成仅1.46万年，但有超过500种明显起源于同一始祖物种的慈鲷鱼生活在那里。这些慈鲷鱼生殖隔离的形成很大程度上可以归因于行为特征的差异和颜色的差别，它们的DNA序列差别很小。在这个群体中，差不多平均1000年就可以产生一个新的物种，但是，维多利亚湖中的其他鱼类并没有这么快的进化速度；通常，形成一个新物种可能需要几万年的时间。

两个近缘种群一旦因一种或更多的杂交繁殖障碍而彼此完全隔离，将会一直彼此独立进化，随着时间的推移，又会产生分化。自然选择是产生这种分化的重要原因。就像前文提到过的加拉帕戈斯雀一样，为了适应不同的生活方式，近缘种通常具有许多不同的构造和行为特征。然而有时，近缘物种间仅有很少的地方明显不同。昆虫常表现出这一点。例如，拟果蝇和毛里求斯果蝇两种果蝇有着非常相似的身体结构，表观上仅雄性生殖器有所不同。然而，它们确确实实是两种不同的物种，相互之间几无交配意愿。与其类似，人们最近发现，常见的欧洲伏翼蝙蝠应该被分为两个物种。这两种蝙蝠在自然条件下并不交配，它们的叫声以及DNA序列也都不同。相反，如同我们之前提到过的那样，有许多属于同一物种但有显著差异的种群间是可以杂交繁殖的。

这些例子共同说明，两种群间可观察的特征上的差别与生殖隔离的强度之间没有绝对的相关关系。两物种间的差异程度，也与距它们之间出现生殖隔离的时间长短没有密切关系。这一点可以通过以下的例子说明：生活在岛屿上的物种，如加拉帕戈斯雀类，虽然只进化了相对较短的时间，不同的物种间差异却很大，而与之相较，南非相近的鸟类经过了较长时间的进化，但其中很多鸟类间的差异却很小（见第四章图13）。类似地，根据化石记录，许多生物千百万年中几乎没有变化，随后急剧转变为新形式，古生物学家通常将它们认定为新物种。

无论是理论模型还是实验室实验都表明，强烈的自然选择可以在100代甚至更短的时间里对物种特征产生深远的影响（见第五章）。例如，为了增加黑腹果蝇一个种群腹部刚毛的数量，对这种果蝇进行了人为选择。80代后，这种选择导致果蝇腹部平均刚毛数量增加了3倍。与之类似，相较于生活在400万年（大概20万代）前类猿的祖先，我们现代人的颅骨平均大小增加了。相反，一旦生活在稳定环境中的生物适应了环境，它们的特征就不会有太大的变化。通常很难从化石记录中分辨出，可见的“急剧”进化改变是否就代表了一个新物种（无法与它的祖先杂交繁殖）的起源，或者仅仅是响应环境的变化而进化出的一个新的世系分支。不管是哪种情况，急速的地质变化都是必须的。

最后，对于发生在许多单细胞生物，比如细菌中的无性繁殖，物种又是怎样定义的？在这里，根据能否杂交繁殖来定义物种毫无意义。为了在这些情况下进行分类，生物学家只是依据主观的相似程度的标准：依据具有实际意义的特征（比如细菌细胞壁的构成）或是更多地依据DNA序列的不同。在进行特征衡量时，那些十分相似的个体聚为一类，人们将之划分为同一物种，反之，其他没有聚成一类的个体，就认为是不同的物种。

## 物种间的分子进化和分化

考虑到两物种互相分离独立进化的时间长短和形态特征的差别大小之间的关系并不规则，生物学家在推断两物种的关系时，越来越多地参考不同物种的DNA序列信息。

就如同类比同一个词在不同但有关联的语言中的拼写一样，人们在不同物种同一基因的序列上也可以发现相同之处和不同之处。例如，英语中的“house”、德语中的“haus”、荷兰语中的“huis”和丹麦语中的“hus”是同一个意思，发音也很相似。这些词之间的不同点有两种。首先，同一位置上的字母不同，如英语的第二个字母是o，而德语则是a。其次，有字母的增加和减少，英语中末尾的e在其他语言中没有出现，丹麦语比德语少了第二个位置上的a。由于缺少更多的有关语言间历史关联的信息，人们很难确切明了这些变化的发展轨迹。尽管人们知道，只有英语的“house”末尾有e的事实强有力地表明了这个e是后来加上去的，而“hus”拼写最短表明了丹麦语中这个词元音的缺失。如果对大量的单词样本进行这样的比较，不同语言的不同点就可以用来衡量它们的关系，这些不同点与语言发生分化的时间有着密切的关联。虽然美式英语从英式英语中分离出来仅有几百年的时间，但是，包括不同方言的发展在内，两者间的分化却很明显。荷兰语和德语有着更大的分化，法语和意大利语间的分化更甚。

同样的规则也适用于DNA序列。在这种情况下，对于那些为蛋白质编码的基因，由DNA单个字母的插入和缺失引起性状改变的情况是很少见的，因为字符的插入或缺失通常会在很大程度上影响蛋白质氨基酸序列，使其功能丧失。近缘物种间，基因的编码序列的变化大多数包括了DNA序列单个字母的变化，比如把G换作了A。图8的例子中列出了人类、黑猩猩、狗、老鼠和猪的促黑激素受体的部分基因序列。

两种不同生物的同一类基因的序列差异表现在DNA字母的数量上，通过比较字母数量，人们可以准确地衡量这两种生物的分化程度，而这种衡量用形态学上的异同是很难做到的。掌握了基因的编码方式，我们就可以知道哪一种差异改变了与所研究的基因相应的蛋白质序列（**置换**改变），哪一种差异没有引起改变（**沉默**改变）。例如，图8中列出的对人类和黑猩猩的促黑激素受体基因序列差异的简单计数，显示出所列的120个DNA字母有四种差异。不同物种的全序列（忽略掉小区域的DNA字母增加和减少）与人类的基因序列相比，不同之处的数量如下表所示。

| 与人类相比 | 相同氨基酸（沉默差异） | 不同氨基酸 |
|---|---|---|
| 黑猩猩 | 17 | 9 |
| 狗 | 134 | 53 |
| 老鼠 | 169 | 63 |
| 猪 | 107 | 56 |

最近的研究表明，人类和黑猩猩的53种非编码DNA序列中有差别的部分占全部字母的0到2.6%，平均值仅为1.24%（人类和大猩猩之间为1.62%）。这一结果解释了为什么人们认为黑猩猩是我们的近亲而大猩猩不是。如果人类跟猩猩比较，差异就更大了，跟狒狒之间的差异更甚。关系更远的哺乳动物，如肉食动物和啮齿类动物，比灵长类动物跟人类在序列水平上的差异更大；哺乳动物与鸟类的差异，比哺乳动物之间的差异更大，诸如此类，不一而足。序列比较所揭示的关系图谱，与根据主要动

植物物种在化石记录中出现的时代作出的推测结果大体一致，这一点也和进化论推理结果一致。

这张显示序列差别的表格表明，沉默改变比置换改变更为常见，即便沉默改变在如黑猩猩和人类这样非常近缘的物种中也十分少见。显然，这是因为大部分蛋白质氨基酸序列的改变会在一定程度上损害蛋白质的功能。就像我们在第五章中提到过的，突变造成的不大的有害作用会引起选择，选择作用很快将突变体从群体中清除出去。因此，大多数引起蛋白质序列变化的突变不会增加物种间累积的基因序列的进化差异。但是，也有越来越可靠的证据表明，一些氨基酸序列的进化由作用于随机有利突变的自然选择所推动，进而引发分子层面的适应性出现。

不同于改变氨基酸的突变常常带来的有害作用，基因序列的沉默改变对生物功能几乎没有影响。由此可以理解，物种间的大部分基因序列差异都是沉默改变。但是，当一个新的沉默改变在种群中出现，它仅仅是相关基因成千上百万个副本中的一个（每种群每个体中有两个副本）。如果一个突变不能为它的携带者带来任何选择优势，这一突变又如何能在种群中传播开来？答案是，有限种群中会出现变异体（遗传漂变）的频率发生偶发性变化的现象，这个概念我们在第二章中作了简单的介绍。

下面介绍这个过程的运作方式。假设我们在对黑腹果蝇的一个种群进行研究。为了种群世代延续，每个成年果蝇平均必须产生两个子代。假设果蝇种群在眼睛的颜色上有差别，某些携带突变基因的个体眼睛是亮红色的，而不携带突变基因的所有其他个体眼睛是通常的暗红色。如果不管哪种个体都能产生同等数量的后代，那么在眼睛的颜色上就不存在自然选择，这种突变的影响就被称为**中性的**。因为这种选择的中性影响，子代从亲代继承的基因是随机的（如图18所示）。有些个体没有后代，而其他个体可能碰巧产生多于两个的后代。因为不论有无携带突变基因，个体产生的子代数量都不可能完全相同，这就意味着，子代中突变基因的频率将不会与亲代一样。因而在多次传代过程中，种群的基因构成会出现持续的随机波动，直到总有一天，或者种群中的所有个体都携带有产生亮红色眼睛的基因，或者这一突变基因从种群中消失、所有个体都带有产生暗红色眼睛的基因。在一个小种群中，遗传漂变的速度很快，不需要多长时间，种群所有的个体都变成一样的了。大种群则需要更长的时间来完成这一过程。

这就解释了遗传漂变产生的两类影响。首先，在一个新的变异体漂变至最后从种群中消失或者该变异体最终在种群中100%出现（**固定下来**）的过程中，受该基因影响的性状在种群中是多变的。突变引入的新中性变异体，以及遗传漂变导致的变异体频率的改变（以及有时发生的变异体基因的消失）决定了种群的多样性。对不同种群个体的同一基因DNA序列的检测结果揭示了这一过程所造成的沉默位点的变化性，这一点我们在第五章中有所阐述。

遗传漂变的第二个影响是，最初非常稀少的、就选择来说属于中性的变异体有机会扩散至整个种群、取代其他变异体，尽管它更有可能从种群中丢失。遗传漂变由此导致两个隔离种群的进化分离，甚至是在没有自然选择作用推动的情况下。这是一个很缓慢的过程。它的速度取决于新的中性突变的出现速度，以及遗传漂变造成基因更迭的速度。值得注意的是，最终两物种DNA序列分化的速率只取决于单个DNA字母突变的速率（亲代字母变异并传给子代的频率）。在这一点上一个直观的解

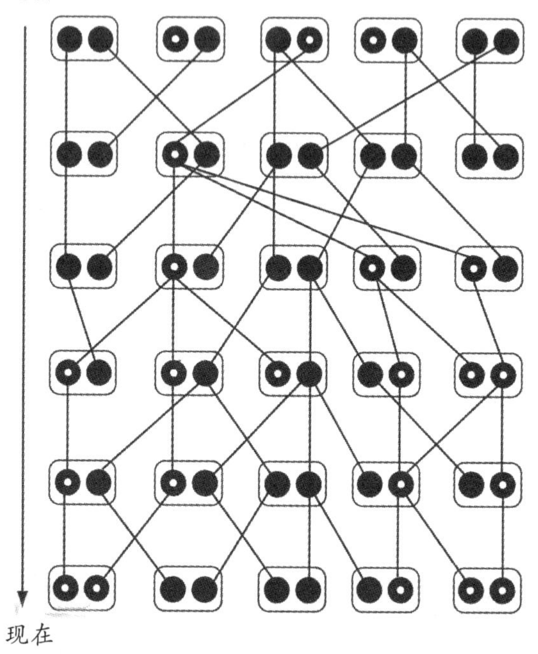


图18 遗传漂变。这张图显示的是一个基因在有五个个体的种群中，经过六个世代中遗传漂变的过程。每个个体（以一个空心的形状表示）有两个该基因的副本，分别来自亲本的一方。个体基因副本的不同DNA序列没有详细列出，而用有或无白点的黑色圆盘来表征。以文中提到过的果蝇为例，白色圆点代表引起亮红色眼睛的变异基因，黑色圆盘代表暗红色眼睛的变异基因。第一代中，有三个个体同时拥有一个白色圆点基因和一个黑色圆盘基因。因此，种群中30%的基因为白色圆点基因。图中显示了每一代的基因谱系（为了方便，假设个体既可作为雄性也可以作为雌性生殖，就像许多如番茄一样雌雄同体的植物、如蚯蚓一样雌雄同体的动物）。出于偶然，一些个体会比其他个体产生更多的后代，而有的个体产生的后代数量则较少，甚至可能没有存活下来的后代（例如，第二代中最右边的个体）。因此，每一世代的白色圆点基因和黑色圆盘基因的副本数量会有波动。第二代仅有一个个体携带有白色圆点基因，到了第三代有三个个体从这个个体遗传了该基因，这就使得该类基因的比例从10%提高到了30%；下一代中是50%，等等。

释是，如果自然选择没有起任何作用，除了序列中突变出现的频率以及从两物种最后一位共同祖先到现在所经历的时间这两点外，就没有什么会影响两物种间突变差异的数量。大种群每一世代可以产生更多的新突变，仅仅是因为可能发生突变的个体数量更多。但是就像上文中阐述的那样，遗传漂变在小种群中会更快发生。结果是，种群规模所造成的两方面影响正好相互抵消，因而突变频率是种群分化速度的决定因素。

这一理论结果对我们判定不同物种间关系的能力有重要启示。这意味着一个基因的中性变化随时间而累积，累积的速度取决于基因的突变速度（分子钟原理，在第三章中有提及，但没有解释）。因而基因序列的变化可能是以一种类似时钟的方式运行，而不是自然选择造成的特征变化。形态变化的速度则高度依赖于环境的变化，并且速度可能有变化，方向可能逆转。

即便是分子钟也不十分精准。同一个世系内的分子进化速度会随时间而变化，不同世系间的也是如此。然而，当没有化石依据的时候，生物学家们可以利用分子钟粗略地估算不同物种分化的时间。为了对分子钟进行校准，我们需要一个分化时间已知的最近缘物种的序列。分子钟最重要的应用之一是确定现代人世系和黑猩猩、大猩猩世系分化的时间，而这一时间没有独立的化石依据。利用包含了大量基因序列的分子钟可以对六七百万年的时间进行较为可信的估算。因为中性序列进化速度取决于突变速度，而DNA单个字母通过突变发生变化的速度非常低，所以分子钟极其慢。人类和黑猩猩的DNA字母之间存在约1%的不同，这一事实与超过10亿年里单个字母只变化一次相契合。这与实验测量得到的突变速率的结果一致。

分子钟也被用于研究蛋白质的氨基酸序列。上文中已经提

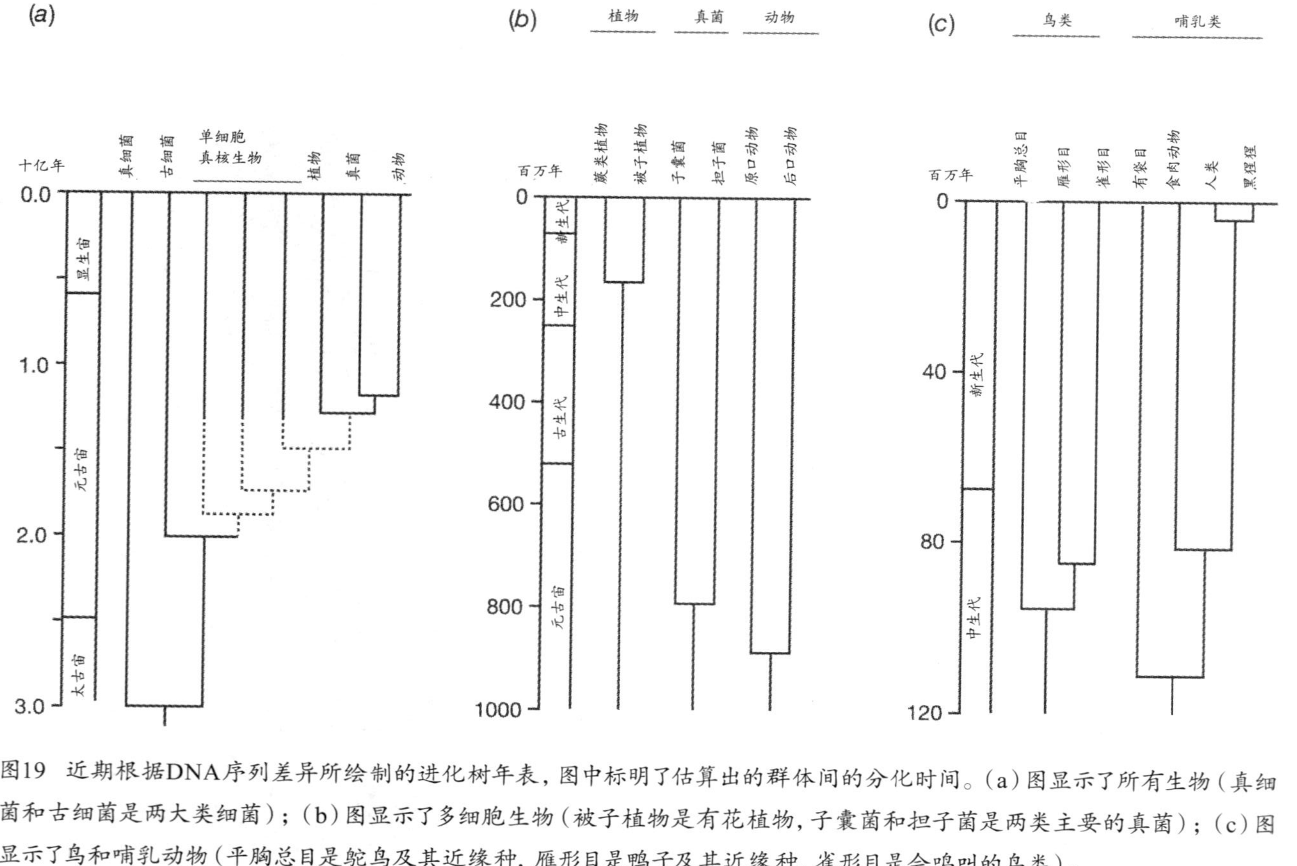


图19　近期根据DNA序列差异所绘制的进化树年表，图中标明了估算出的群体间的分化时间。(a)图显示了所有生物（真细菌和古细菌是两大类细菌）；(b)图显示了多细胞生物（被子植物是有花植物，子囊菌和担子菌是两类主要的真菌）；(c)图显示了鸟和哺乳动物（平胸总目是鸵鸟及其近缘种，雁形目是鸭子及其近缘种，雀形目是会鸣叫的鸟类）。

到过，蛋白质序列进化慢于沉默DNA差异的进化，因而有助于完成一个棘手任务：对分化了很长时间的物种进行比较。在这类物种之间，大量的变化发生在DNA序列的某些位点，因此不可能准确计算出发生突变的数量。致力于重建现有生物主要群体间分化时间的科学家，于是采用了来自缓慢进化着的分子的数据（图19）。当然这样的数据只是粗略的估算，但是，通过对多个不同基因的估算累积，可以提高计算过程的准确性。通过对以不同速度进化的基因序列信息的审慎使用，进化生物学家能够绘制某些生物群体间的关系图谱，这些生物的最后一位共同祖先生活在距今10亿年前甚至更久远。换言之，我们近乎重建了生命世系树。

第七章
# 一些难题

随着进化论变得越来越被人们理解，以及不断被生物学家证实，新的问题出现了。不是所有的问题都已经解决了，争论始终存在于新老问题之中。在本章中，我们将描述一些表面上很难解释的生物现象，其中一些达尔文自己已经解决，剩下的则成为了之后研究的主题。

## 复杂的适应性是如何进化的？

自然选择进化论的批评者经常提及从蛋白分子到单个细胞再进化到眼睛和大脑等复杂生物结构的困难性。一个只通过自然选择产生的功能齐全且具有良好适应性的生物机器如何能依靠偶然的突变来运行？理解这些何以发生的关键表现在“适应”这个词的另一层含义中。生物体以及它们复杂的结构的进化过程，就像工程师造机器一样，许多方面都是先前结构的修改（适应）版本。在制造复杂的机器和设备时，最初不那么完美的版本随着时间推移不断精炼，增加了（适应）新的甚至是完全想象不到的用途。全膝关节置换术的发展过程就是一个很好的例子：一个粗糙的初始方案足以解决问题，但是被不断改造得越来越好。与在生物进化中类似，按今天的标准来看，许多早期设计的改变看上去很小，但是每一个变化都是在前一个基础上的改进，而且可以被膝外科医生所利用，每一个过程都在复杂的现代人工膝盖的发展中起了作用。

“设计”被不断修改完善的过程就像是在大雾天爬山。即使没有一定要登顶的目的（或者甚至不知道山顶在哪），但只要遵循一个简单的原则——每一步都向上——就会离山顶（至少局部的顶点）越来越近。用某种方式简单地使其中一个构件更好地工作，那么即使没有设计师，整个设计最后也有了改进。在工程上，改进的设计通常是不同工程师对机器改进做多重贡献的结果，最早的汽车设计者看到现代汽车可能会大吃一惊。在自然进化中，改进来自对生物体所谓的“修补”，很小的变化就使这个生物体能更好地生存或繁殖。在一个复杂结构的进化中，许多不同的性状必然是同时进化的，这样结构的不同部分才可以很好地适应一个整体的功能。我们在第五章中看到，相对于主要的进化演变所用的时间，有利的性状，即使它们最初很罕见，也可以在短时间内在种群中扩散。一个已经运作但可以改进的结构体发生连续的小的改变，因此可以产生大的进化演变。在经历数千年后，可以想象到即使一个复杂结构也会发生彻底改变。足够长的时间以后，此结构将在许多不同方面与原始状态相异，于是后代个体可以拥有祖先从未有过的组合性状，就像现代汽车与最早的汽车有很多不同点一样。这不仅是一个理论上的可能性：就像在第五章所描述的，动物和植物育种人经常通过人工选择来实现这一可能。因此我们不难了解到，这些由许多彼此协调的部件所组成的性状是如何由自然选择所引发的。

蛋白质分子的进化有时被认为是一个特别困难的问题。蛋白质结构复杂，每个部位必须相互作用才能正常运转（许多蛋白质必须也和其他蛋白质和分子（有些情形下包括DNA）相互作用)。进化论必须能够解释蛋白质的进化。蛋白质一共有20种不同的氨基酸,因此在一个100个氨基酸长度的蛋白质分子中(比许多实际的蛋白质分子短),正确的氨基酸出现在特定位置的概率是1/20。如果100个氨基酸随机混合在一起,那么序列中每一个位置都有正确的氨基酸、形成一个正常的蛋白质的概率显然非常小。因此有人声称,组装一个发挥功能的蛋白质和用龙卷风吹过废品场来组装一架大型客机的概率差不多。一个发挥功能的蛋白质确实不能通过在序列中每个位置随机挑选一个氨基酸组装而成,但是,就像上文中的解释所阐明的,自然选择不是这样工作的。蛋白质最初可能只是几个氨基酸的短链分子,这样可以使反应快一点,接着在进化中不断得到改进。我们可以忽略数百万不发挥作用的潜在的非功能序列,只要蛋白质序列在进化过程中能比没有蛋白质的时候为反应提供更好的催化作用,然后不断地随着进化时间持续改进。我们很容易在总体上了解到,连续的变化(序列的改变或增长)将如何改进一个蛋白质。

关于蛋白质工作原理的研究成果支持了上述观点。蛋白质中对其化学活动至关重要的部分通常都是序列里非常小的一部分;一个典型的酶只有一部分氨基酸与化学物质发挥作用然后改变这种化学物质,其他蛋白链大部分只是简单地提供支架以支撑参与这个作用的部分的结构。这说明一个蛋白质要发挥功能关键只取决于小部分氨基酸,因此蛋白质序列的一些很小数量的变化就可以进化出一个新的功能。许多实验都证实了这一点,这些实验通过人工诱导蛋白质序列的变化来使它们适应新的功能。我们已经证明,通过这些方式(有时仅仅需要改变一个氨基酸)可以很容易给蛋白质的生物活动带来剧烈的变化,在自然进化的变化中也有类似的例子。

类似的答案也可用以回答连续酶反应的路径如何进化，例如那些产生生物体必需的化学物质的酶（见第三章）。有人可能认为，即使最终产物是有用的，由于进化没有先见之明，无法建立起一个功能完全的连锁酶反应，因此也不可能进化出这些路径。这个谜语的答案是显而易见的。在早期生物的环境中可能存在许多有用的化学物质，随着生命不断进化，这些物质变得稀有了。能够将一种类似化学物质变成有用化学物质的生物体将会受益，于是酶就会进化以催化这些变化。这时有用的化学物质就可以用相关的物质合成，因此一个有前身和产物的短的生物合成路径将会受到青睐。通过这样连续的步骤，从它们的最终产物往后推，就可以进化出能够为生物体提供必需的化学物质的路径。

如果复杂的适应，就像进化生物学家提出的那样，真的是逐步进化的，那么我们应该可以找到这些性状进化的中间阶段的证据。这样的证据的来源有两种：化石记录里中间过程的发现，以及具有介于简单和先进状态之间的过渡特征的现存物种。在第四章，我们描述了连接各迥异形式的中间化石，这些化石支持了进化是逐步改变的理论。当然，在很多时候我们无法找到进化的中间生物，尤其是当我们追溯到更为遥远的年代时。特别是多细胞动物的主要成员，包括软体动物、节肢动物和脊椎动物，几乎全部都突然出现在寒武纪（5亿多年前），而且几乎没有关于它们祖先的化石证据。关于它们之间的关系，最新的DNA研究有力地证明，早在寒武纪之前这些群体就已经是独立的世系（图19），但是我们没有任何关于它们形象的信息，可能是因为它们是软体而不可能变成化石。但是不完全的化石记录并不意味着中间阶段不存在。新的中间阶段的证据正在持续被发现，距今最近的是在中国发现的一块1.25亿年前的哺乳动物化石，它与现代胎盘哺乳动物特征类似，但是比这个物种之前所知的最古老的化石早了4000万年。

另一种类型的证据来源于习性的比较，这是我们研究那些没有变成化石的特征的唯一途径。就像达尔文在《物种起源》一书的第六章中指出的那样，一个简单但令人信服的例子是飞行。没有化石连接着蝙蝠和其他哺乳动物，在沉积物中发现的距今6000多万年的第一批蝙蝠化石，和现代蝙蝠一样有高度改良的四肢。但是有例子证实一些现代哺乳动物具有滑翔能力却不会飞。最为人类所熟悉的是鼯鼠，它们与普通松鼠很相似，除了连接前后肢的翼膜。翼膜就像粗糙的翅膀，可以使鼯鼠荡起来的时候滑行一段距离。在其他哺乳动物——包括所谓的飞狐猴（并不是真正的狐猴，和鼯鼠也无关）——和蜜袋鼯中，相似的滑翔适应性已经独立进化。蜥蜴、蛇和青蛙的滑行物种也是我们所知道的。很容易想象滑行能力可以减少树上生活的动物被捕食者抓到或吃掉的风险，因此在树枝间跳来跳去的动物的身体逐步改变而进化出滑行能力。滑行所用到的皮肤区域逐渐增大，前肢发生改变以适应这种增大，这些都将有利于生存。飞狐猴有一张大得可以从头部延伸到尾巴的膜，尽管只能滑行不能飞，但和蝙蝠的翅膀很相似。一旦可以高效滑行的翅膀结构进化形成，翅膀肌肉组织发育并产生动力就不难设想了。

作为另一个复杂适应的例证，达尔文也研究过眼睛的进化。脊椎动物的眼睛是一个高度复杂的结构，在视网膜上有感光细胞，透明的角膜和晶状体使得图像可以在视网膜上聚焦，还有可以调整焦距的肌肉。所有脊椎动物的眼睛构造都基本相同，但是在细节上有许多变化以适应不同的生活模式。没有视网膜，晶状体似乎毫无作用，反之亦然，那么这样一种复杂的器官是怎么进化出来的？答案是没有晶状体时视网膜绝不是无用的。许多种无脊椎动物都拥有不含晶状体的简单的眼睛，这些动物不需要看得清楚，眼睛可以感知明暗进而察觉捕食者就足够了。事实上，在不同的动物中可以看到简单的感光受体和各种类型可以产生图像的复杂装置之间的一系列中间形态（图20）。甚至单细胞真核生物，也能通过由一群光敏蛋白视紫红质分子组成的受体感知并回应光。所有动物的眼睛中都含有视紫红质，在细菌中也能发现。从细胞的这种可以感知光的简单能力开始，不难想象聚光能力将逐步进化提高，最终成为一个可以聚焦并产生清晰图像的晶状体。正如达尔文所说：

> 在活体中，变异会引起轻微的改变，……自然选择将娴熟地挑选每一个改进。让这个过程持续数百万年，在每一年中作用在许多种类的数百万的个体上；我们有理由相信，一个比玻璃更好的、活的光学器官将由此产生。



## 我们为什么会衰老？

作为一个整体，年轻的身体令人惊叹，就像眼睛，是近乎完美的生物机械。但这种“近乎完美”有一个反面的问题，就是它们在生命中持续的时间不长。为什么进化会允许衰老发生？近乎完美的生物由于衰老而变成自己微弱的影子是诗人们钟爱的主题，尤其是当他们预见这些将发生在爱人身上：

> 于是我不禁为你的朱颜焦虑：

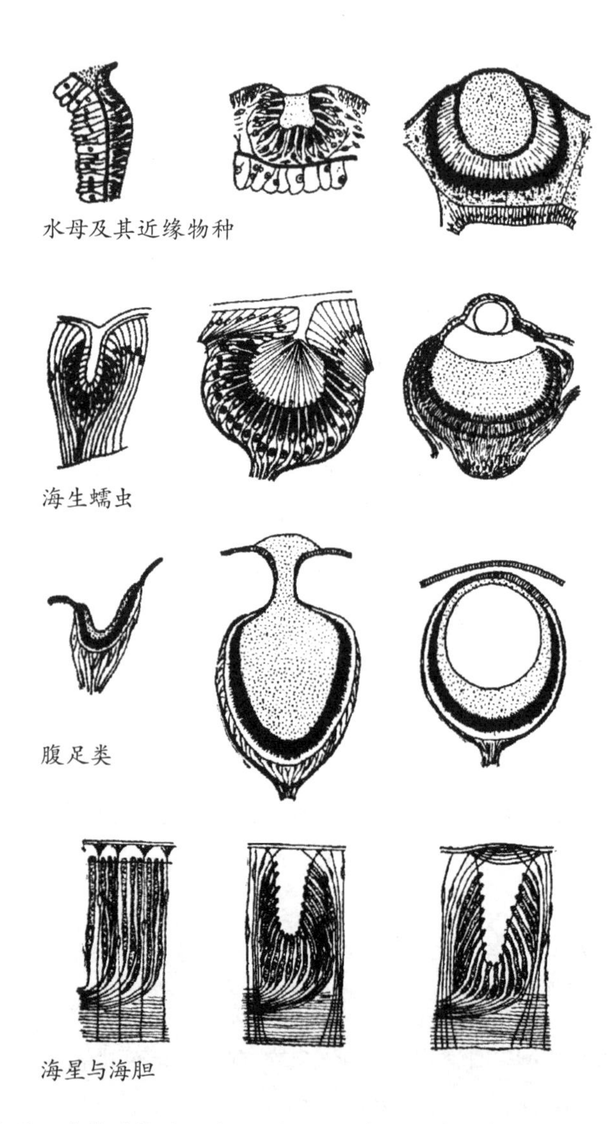


图20 各种无脊椎动物的眼睛。从左至右，每一行展示着给定的类别中不同物种由低到高的眼睛类型。例如海生蠕虫（第二行），左边的眼睛只由一些光敏和色素细胞组成，有一个透明的圆锥体投射到它们中间。中间的眼睛有一个充满了透明的胶状物的腔体和有大量感光细胞的视网膜。右边的眼睛在腔体前有一个球面透镜和更多的光受体。

终有天你要加入时光的废堆，
既然美和芳菲都把自己抛弃，
眼看着别人生长自己却枯萎；
没什么抵挡得住时光的毒手，
除了生育，当他来要把你拘走。①

——莎士比亚的“十四行诗12”

衰老当然不局限于人类，在几乎所有的植物或动物中都可以观察到。为了测定衰老程度，我们可以研究许多置于保护区域中的个体，也就是排除“外部”原因比如捕食造成的死亡的环境，这样生物可以比在自然环境中所活的时间长很多。一直追踪下去，我们可以测定不同年龄段的死亡率。即使在被保护的环境中，刚出生的个体的死亡率通常也较高；随着个体长大，死亡率会下降，但成年期之后再次上升。在大多数得到细致研究的物种中，成年个体的死亡率随着年龄的增长而稳步上升。然而不同物种的死亡率的模式差别较大。相比人类这样个体大且寿命长的物种，个体小且寿命短的生物，例如老鼠，在相对年轻时的死亡率高得多。

衰老导致死亡率的上升，这是生物体的多项功能随着年龄增加而退化的结果：似乎所有的东西都在变糟，从肌肉力量到精神能力。在多细胞生物中几乎普遍发生的老化（这看似一种退化）与自然选择导致适应性进化的观点矛盾，这可能看上去是进化论的一个严峻的困境。一种回答是适应性绝不是完美的。在生物体生存所必需的系统中，长期以来累积的损害将不可避

① 梁宗岱译文。

免地避免衰老，而自然选择可能根本无法阻止它发生。实际上，复杂的机器例如汽车的年均故障次数，也会随着时间增加而增加，这与生物体的死亡率非常类似。

但这不可能是答案的全部。单细胞生物体例如细菌简单地通过分裂生成子细胞来繁殖，通过这些分裂产生细胞谱系已经持续了数十亿年。它们不衰老，但是持续分解已损坏的组分并用新的替换掉。它们可以无限繁殖下去，前提是自然选择清除了有害的突变。对一些生物（例如果蝇）人工培养的细胞而言，这也是可能的。多细胞生物的生殖细胞谱系也可以在每一代延续，所以为什么整个生物体不会维持修复过程？为什么我们大部分的身体系统显示出某种衰老？例如，哺乳动物的牙齿随着年龄增长而磨损，最终导致饿死在自然界。这不是必然的，爬行动物的牙齿可以不断更新。不同物种的不同衰老速度展示了不同效果的修复过程和随着年龄增长的保持程度：一只老鼠最多能活3年，然而一个人可以活超过80年。这些物种差异表明衰老过程也是进化的，因此衰老需要一个进化论的解释。

在第五章中，我们看到对多细胞生物的自然选择通过个体对后代贡献的差异发挥作用，包括它们产生的后代的数量以及生存机会的差异。此外，所有个体都有事故、疾病和捕食导致死亡的风险。即便这些死因发生的概率与年龄无关，生存几率也随着年龄的增加而下降，我们和汽车都是这样：如果从第一年到下一年良好运转的概率是90%，5年后这个概率是60%，但是50年后就只有0.5%。因此对生存和繁殖的自然选择倾向于在生命的早期而不是晚期进行，仅仅因为平均来看更多的个体能够活着感受其有益的效果。事故、疾病和捕食造成的死亡率越高，自然选择将越强烈地倾向于生命早期的改进，因为如果这些外部原因造成的死亡率很高，那么很少有个体可以存活到老年。

这个观点表明衰老进化是由于与后期的变异相比，在生命早期有更多的有利于生存或繁殖的备选变异。这个概念类似于我们熟悉的人寿保险：如果你年轻，那么购买一定数量的保险将花费少，因为你可能有许多年是提前支付。自然选择引起衰老的作用途径主要有两种。上面提到的观点表明，有害的突变如果发生在生命早期，将遭受最强烈的抵抗。选择能引起衰老的第一种方法是保持种群中很少发生早期突变，同时允许在生命晚期发生变得普遍。许多常见的人类基因疾病确实是由那些在晚年出现有害作用的突变导致的，例如阿尔兹海默症。第二种途径是，在生命早期带来有利影响的变异体比只在晚期带来有利影响的更有可能在整个种群中扩散。生命早期的改进可以进化，即使这些好处是以之后有害的副作用为代价。例如，年轻时高水平的生殖激素可以提高妇女的生育能力，但是之后却有患乳腺癌和卵巢癌的风险。实验结果证实了这些预测。例如，通过只用非常老的个体来繁殖，可以保持黑腹果蝇的种群。在几代之后，这些种群进化出更慢的老龄化，但代价是生命早期的生殖成功率降低了。

衰老的进化理论预测，外因死亡率较低的物种应该比更高的物种衰老得慢并且寿命更长。躯体大小和衰老速度之间确实有很强的联系，躯体较小的动物比大的衰老得快得多，而且生殖时间要早。这可能是由于许多小动物更容易遭受意外事故伤害或被捕食。当我们考虑到被捕食的风险时，具有相似尺寸但野外死亡率不同的动物所具有的迥异的衰老速度往往就容易理解。许多飞行生物以长寿著称是有道理的，因为飞行是对捕食者很好的防御。一个相当小的生物，例如鹦鹉，可以比一个人的寿命更长。蝙蝠与同等体重的陆生哺乳动物例如老鼠相比，寿命要长得多。

我们人类自己也可能是进化学中老化速度较慢的一个例证。我们的近亲黑猩猩即使在人工饲养环境下也很少有活过50年的，并且比人类更早地繁殖，平均生殖年龄为11岁。因此，从人类和猿偏离我们共同的祖先开始，人类可能大幅降低了老化速度并推迟生殖成熟期。这些改变可能是由于提升的智力和合作能力，它们减少了外部死亡原因的威胁和早期生育的优势。早期和晚期生育相对优势的改变可以在当今社会中发现甚至量化。人口普查数据明确表明，工业化导致成年人的死亡率急剧下降，这改变了自然选择对人类老化过程的影响。不妨考虑一下由罕见的基因突变引起的亨廷顿氏病（退化的大脑失调），这种病发病较晚（在30岁或者更晚）。在由于疾病和营养不良而死亡率较高的人群中，很少有个体能活到40岁，亨廷顿氏病患者的后代数量比其他人平均只略少（少9%）。在工业化社会，死亡率很低，人们经常在这种疾病可能发生的年龄才有小孩，结果患病者比不患病者平均少了15%的后代。如果目前的状况持续，自然选择会逐步降低在育龄后期起作用的突变基因的出现频率，老年人的生存率将会提高。罕见但影响重大的基因，例如亨廷顿氏病，对人群整体的影响较小，但是其他许多部分由基因控制的疾病对中老年人的影响很大，包括心脏病和癌症。我们可能希望这些基因的发生率因为这种自然选择而随着时间下降。如果工业化社会中低死亡率的特点持续几个世纪（一个大胆的假设），那么将会出现缓慢但是稳定的基因改变，降低衰老速率。

## 不育的社会性昆虫的进化

进化论的另一个问题是许多类型的社会性动物中存在的不育个体。在社会性的黄蜂、蜜蜂和蚂蚁中，巢穴中有一些不生殖的雌性个体，即工蚁或工蜂。生殖的雌性是群体中的极少数（通常只有一个蚁后或蜂后）；雌性工蚁或工蜂照看蚁后或蜂后的后代并维护和建造巢穴。另一种主要类型的社会性昆虫——白蚁，雄性和雌性都可以成为工蚁。在高级的社会性昆虫中，通常有几种不同的"阶级"，它们扮演不同的角色，通过行为、大小和身体结构的不同来区分（图21）。

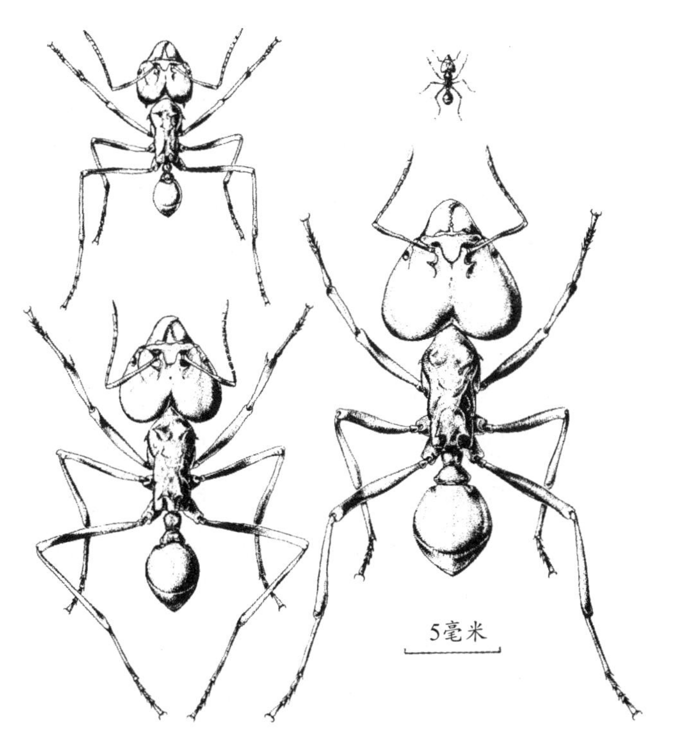


图21 来自同一个群体的切叶蚁属的工蚁阶级。右上角最小的工蚁负责照看切叶蚁耕作的真菌花园，大的则是负责保卫蚁巢的兵蚁。

一个了不起的新发现是，社会性巢居哺乳动物中的几个物种与这些昆虫有类似的社会结构，巢穴中大部分的居住者是不育的。最为人熟知的是裸鼢鼠，非洲南部沙漠地区的一种穴居啮齿类动物。巢穴中可能居住着几十个成员，但是只有一个生殖的雌性，如果她死了，其他的雌性通过战斗决出胜利者来取代她。拥有不育劳作成员的社会性动物的系统因此进化成完全不同的动物种群。这些物种给自然选择理论带来一个显而易见的问题：为什么会进化出失去生殖能力的个体？既然劳作成员自己不生育，因此无法直接接受自然选择，那么它们对分工中的专业角色的极端适应是如何进化来的？

达尔文在《物种起源》一书中提及了这些问题，而且回答了部分。答案就是：社会性动物的成员通常都是近亲，例如裸鼢鼠或者蚂蚁，经常共有同一个母亲和父亲。基因变异体导致某携带者放弃自己的繁殖机会而养育它的亲属，这可能有助于将近亲的基因传递给下一代，近亲的基因通常（由于亲缘关系）与施助个体的基因相同（对于一对姐弟或兄妹，如果其中一个从父母那里遗传了一个基因变异体，那么另外一个也有此变异体的概率是50%）。如果不能生育的个体的这种牺牲导致成功生存和繁殖的近亲数目增加，那么这种“劳作基因副本”数目的增加可以超过由于它们自身不能繁殖而导致的数目减少。关系越近，弥补损失所需的量将越少。J.B.S.霍尔丹曾经说过：“为了两位兄弟或八位表亲我愿献出我的生命。”

**亲缘选择**理论为理解社会性动物中不育的起源提供了架构，现代研究表明它可以解释动物社会的许多细节，包括那些不像不育昆虫阶级那么极端的特征。例如，在一些鸟类中，幼年雄性没有试图去交配，而是当年幼的兄弟姐妹需要照顾时，在它们父母的巢中扮演着帮手的角色。与之类似，豺狗会在其他成员外出捕猎时照顾年幼的个体。

昆虫不育劳作阶级内部的差异是如何出现的，这个问题与上面所提到的问题略有不同，但是它的答案与上文有一定相关性。特定劳作者的发育受环境信号的控制，例如一个幼虫所得到的食物的数量和质量。然而，对这些环境信号的反应能力却是基因决定的。一个特定的基因变异体可能让蚂蚁中一个不育成员发展成（比如说）兵蚁（下颚比平常的工蚁大）而不是工蚁。如果有兵蚁的群体可以更好地抵御敌人，并且带有这种变异体的群体平均可以繁殖更多，那么这个变异体将提高群体的成功率。如果群体中繁殖活跃的成员是工蚁的近亲，这个引起一些工蚁变成兵蚁的基因变异体将通过蚁后和雄性建立新的群体而传播开来。自然选择因此可以提高这个变异体在该物种的群体中出现的概率。

这些观点同样解释了从单细胞祖先到多细胞生物的进化过程。卵子和精子结合而产生的细胞相互间保持联系，其中大部分丧失了成为生殖细胞和直接为下一代贡献的能力。由于所有相关细胞的基因都是相同的，与另一边的单细胞生物相比，充分提高一部分相关细胞群的生存和繁殖能力将是有利的。非生殖细胞为了整体细胞的利益而“牺牲”了自己的繁殖，有些在发育过程中随着组织的形成和溶解注定要死亡，大部分细胞失去了分裂的潜能，就像我们在讨论衰老的进化时解释的那样。当细胞无视器官的需求而恢复分裂能力时，给生物体造成的严重后果表现为癌症。细胞在发育过程中分化成不同的类型类似于社会性昆虫分化成不同的等级。

## 活细胞的起源与人类意识的起源

进化论中，在生命发展史的两个极端上，存在另外两个重要但是很大程度上未解决的问题：活细胞基本特征的起源和人类意识的起源。与我们刚讨论的问题相比，它们是生命史上独特的事件。它们的独特性意味着我们不能利用对现存物种的比较来可靠地推论他们是如何发生的。此外，关于生命极早期历史和人类行为的化石记录的缺失，意味着我们没有关于进化事件发生次序的直接信息。这当然不能阻止我们猜测这些可能是什么，但是这样的猜测不能用我们已描述过的其他进化问题的解决方法来检验。

对于生命的起源，大部分当前研究的目的是找到类似于地球早期普遍条件的情形，这种条件允许能够自我复制的分子的纯化学聚合，正如我们细胞中的DNA在细胞分裂时的自我复制。这种自我复制的分子一旦形成，不难想象不同类型的分子之间的竞争将进化出能更精确和更快速复制的分子，这就是自然选择作用对它们的改进。通过将简单分子的溶液（早期地球上的海洋可能的存在形态）置于电火花和紫外线下照射，化学家们成功地合成了组成生命的基本化学成分（糖、脂肪、氨基酸以及DNA和RNA成分）。不过这些成分如何组装成类似RNA或DNA的复杂分子，这方面研究进展很有限，而如何使这样的分子自我复制方面的研究进展则更有限，所以我们还远未达到预期的目标（但一直在持续进步）。进一步来说，一旦这个目标实现了，还有一个问题必须解决：如何进化出一个原始的基因编码，使短链RNA或DNA序列能够决定一个简单的蛋白质链。虽然已经有许多的想法，但是至今仍没有能解决问题的明确方法。

类似地，对于人类意识的进化我们也只能猜测。我们甚至很难清晰地表述这个问题的本质，因为众所周知意识是很难准确定义的。大部分人认为刚出生的婴儿没有意识，但很少有人质疑一个两岁的小孩是有意识的。动物在多大程度上有意识也存在激烈的争论，但是喜爱宠物者清楚地认识到狗和猫对主人的意愿和情绪的反应能力。宠物甚至似乎可以操纵主人去做它们意愿中的事情。因此意识可能是一个程度问题，而不是本质，因此原则上很容易想象我们祖先在进化过程中逐步强化自我意识和沟通能力。一些人会认为语言能力是拥有真正意识的最有力的判断准则，即使这种能力在婴儿时以惊人的速度逐步发展。进一步来说，有明显的迹象表明动物具有基本的语言能力，例如鹦鹉和黑猩猩，它们可以通过学习交流简单的信息。我们人类与其他高等动物之间实际上的差距没有表面上那么明显。

虽然对于那些推动了人类心理与语言能力（显然远超过其他任何动物）进化的选择性力量的细节，我们一无所知，但是从进化的角度来解释它们却没有任何特别的神秘之处。生物学家在认识大脑的功能方面正取得飞速的进展，毫无疑问心理活动的所有形式都可以用大脑中神经细胞的活动来解释。这些活动一定受到具体调控大脑发育和运转的基因的控制；像其他的基因一样，这些基因也很容易突变，引起自然选择可以发挥作用的变异。某些基因的突变导致其携带者说话语法方面存在缺陷，于是人们能够识别出与语法控制相关的基因，这一技术已不再是纯粹的假想。如今，我们甚至已经弄清引起相关差异的DNA序列中的突变。

第八章

# 后记

距达尔文和华莱士第一次将他们的想法公之于众已经140年了，我们对进化了解了多少？正如我们已经看到的，现代进化观点在许多方面和他们的十分接近，两者都认为自然选择是引导结构、功能和行为进化的主要动力。主要的不同点在于，由于在两方面的进步，相比于20世纪初，人们现在更加相信在自然选择作用下遗传物质的随机突变引发的进化过程。首先，我们有更丰富的数据，展现了在生物组织中，从蛋白分子到复杂的行为模式，每一个水平层面上自然选择所发挥的作用。其次，我们现在已经理解了对达尔文和华莱士来说还是一个谜的遗传机制。我们现在详细地掌握了遗传的许多重要方面，从遗传信息是如何储存在DNA中的，到这些信息又是如何以特定的蛋白质为中间体、通过调节它们的产生水平来控制生物体性状的。此外，我们现在还知道DNA序列的许多变化几乎不会影响生物体的功能，因此序列的进化改变可以通过遗传漂变的随机过程实现。DNA测序技术使我们能够研究遗传物质本身的变异和进化，也能够通过序列的差异重建物种间的遗传谱系。这些遗传知识，以及我们对自然选择驱动生物体物理和行为特征进化的理解，并不意味着能够对这些特征的所有方面做出严格的遗传解释。基因只规定了生物体能够显现出来的那部分特征的可能范围，实际表达出来的特征常依赖于生物体所处的特定环境。对于高等动物，学习在行为活动中起重要作用，但是动物可以学习的行为范围受限于它的大脑结构，而大脑结构又受限于遗传构成。这一点当然也在跨物种的情况下适用：狗永远也学不会说人话（人也不能嗅到远处兔子的气味）。在人类之中，有强有力的证据表明遗传和环境因素都是引发心理特征差异的诱因；如果人类不遵循其他动物所遵循的这一规律，那才令人吃惊。人类的多数变异都存在于同一个区域群体的个体间，不同群体间的差异则少得多。因此，种族是同质的、彼此独立的存在这种想法是毫无道理的，而某个种族具有遗传上的"优越性"这一说法更是无稽之谈。这是一个科学如何为人们在社会和道德问题上提供决策信息的案例，尽管科学无法直接做出那些决定。

那些我们认为基本上为人类所具有的特征，比如说话的能力、象征性思维的能力以及指引家庭和社会关系的情感，必定反映了始于数千万年前的漫长的自然选择过程，从那个时候起，我们的祖先开始了社会群体生活。我们在第七章中讲到，以社会性群体而居的动物能够进化出非完全自私的行为模式，即不会牺牲其他个体以使自己生命延续或繁殖成功。人们很容易认为，这种特征作为一种对他人的公平感，形成了我们身为社会性动物的进化遗传的一部分，就像亲代对子代的抚育无疑代表了同许多其他动物类似的进化行为。我们再次强调这并不意味着人类行为的所有细节都是受遗传控制的，或是显示了可提高人类适应性的特征。而且，对人类的行为做出的进化上的解释，是很难加以严格测试的。

在进化过程中有进步吗？答案是有保留的"是的"。复杂的动植物都是由不太复杂的动植物进化而来的，生命的历史也展示出从简单的原核单细胞生物体到鸟和哺乳动物的一般进步过程。但是自然选择进化论并没有暗示这一过程是不可避免的，细菌显然还是最丰富和最成功的生命形式之一。这就像是保存了虽然老旧但是仍然有用的工具，比如说现代世界中电脑旁边的锤子。复杂性会随进化下降的例子有很多，比如，穴居物种失去视力，或者寄生虫缺少独立生存所需要的结构和功能。就像我们已经多次强调的那样，自然选择不能预测未来，只能积累在普遍环境中有利的变异体。复杂性的提高可能常会带来更好的功能，就像眼睛，然后这一功能被选中留存。如果这一功能不再与适应性有关，相关结构的退化就在情理之中。

进化也是冷酷无情的。自然选择发挥作用，打磨捕食者的捕猎技巧和武器，不管不顾猎物的感觉。它让寄生虫进化出入侵宿主的精妙装置，即使这会引发强烈的痛苦。它引起衰老。自然选择甚至能让一个物种进化出低生育率，当环境恶化时，该物种就会走向灭亡。然而，化石记录和如今惊人丰富的物种所揭示的生命历史，让我们对30多亿年的进化结果感到惊叹，尽管这都是"自然之战、饥饿和死亡"（达尔文语）的结果。对进化的了解让我们知道了我们在自然界中的真正位置——我们是由冷酷的进化力量所造就的数量极多的生命形式的一部分。这些进化的力量已经给了我们这个物种独特的推理能力，因此我们可以运用远见去减轻"自然之战"。我们应该敬畏进化所造就的东西，保护它们不因我们的贪婪和愚蠢而遭受毁灭，并为我们的后代留存它们。如果我们不去这么做，和其他许多美妙的生物一起，我们自己也会走向灭绝。

# 索 引

（条目后的数字为原文页码）

## A



## B

## C

## D

## E

## F



## G

## H

## I



## J

## K



## L



## M

## N

## O



## P

## R

## S



## T

## U



## V



## W

## Y

Brian and Deborah Charlesworth

# EVOLUTION

## A Very Short Introduction

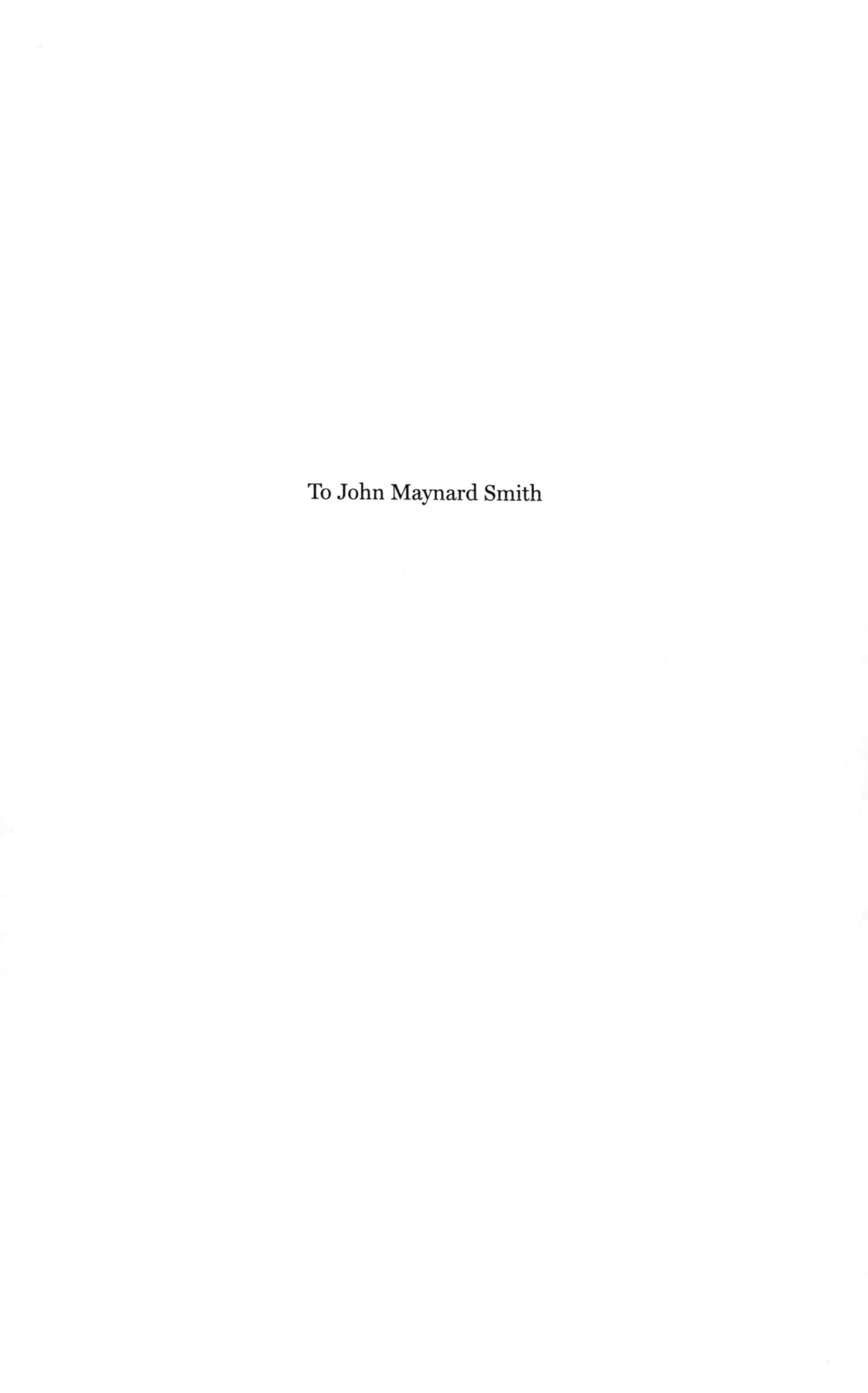

To John Maynard Smith

# Acknowledgements

We thank Shelley Cox and Emma Simmons of Oxford University Press for respectively suggesting that we write this book and for editing it. We also thank Helen Borthwick, Jane Charlesworth, and John Maynard Smith for reading and commenting on the first draft of the manuscript. All remaining errors are, of course, our fault.

# Contents

# List of illustrations

The publisher and the author apologize for any errors or omissions in the above list. If contacted they will be pleased to rectify these at the earliest opportunity.

# Chapter 1
# Introduction

> We are all one with creeping things;
> And apes and men
> Blood-brethren.
>
> From 'Drinking Song' by Thomas Hardy

The consensus among the scientific community is that the Earth is a planet orbiting a fairly typical star, one of many billions of stars in a galaxy among billions of galaxies in an expanding universe of enormous size, which originated about 14 billion years ago. The Earth itself formed as the result of a process of gravitational condensation of dust and gas, which also generated the Sun and other planets of the solar system, about 4.6 billion years ago. All present-day living organisms are the descendants of self-replicating molecules that were formed by purely chemical means, more than 3.5 billion years ago. The successive forms of life have been produced by the process of 'descent with modification', as Darwin called it, and are related to each other by a branching genealogy, the tree of life. We human beings are most closely related to chimpanzees and gorillas, with whom we shared a common ancestor 6 to 7 million years ago. The mammals, the group to which we belong, shared a common ancestor with living species of reptiles about 300 million years ago. All vertebrates (mammals, birds, reptiles, amphibia, fishes) trace their ancestry back to a small fish-like creature that lacked a backbone, which lived over 500 million years ago. Further back in time, it becomes increasingly difficult to discern the relationships between the major groups of animals, plants, and microbes, but, as we shall see, there are clear signs in their genetic material of common ancestry.

Less than 450 years ago, all European scholars believed that the Earth was the centre of a universe of at most a few million miles in extent, and that the planets, Sun, and stars all rotated around this centre. Less than 250 years ago, they believed that the universe was created in essentially its present state about 6,000 years ago, although by then the Earth was known to orbit the Sun like other planets, and a much larger size of the universe was widely accepted. Less than 150 years ago, the view that the present state of the Earth is the product of at least tens of millions of years of geological change was prevalent among scientists, but the special creation by God of living species was still the dominant belief.



The relentless application of the scientific method of inference from experiment and observation, without reference to religious or governmental authority, has completely transformed our view of our origins and relation to the universe, in less than 500 years. In addition to the intrinsic fascination of the view of the world opened up by science, this has had an enormous impact on philosophy and religion. The findings of science imply that human beings are the product of impersonal forces, and that the habitable world forms a minute part of a universe of immense size and duration. Whatever the religious or philosophical beliefs of individual scientists, the whole programme of scientific research is founded on the assumption that the universe can be understood on such a basis.

Few would dispute that this programme has been spectacularly successful, particularly in the 20th century, which saw such terrible events in human affairs. The influence of science may have indirectly contributed to these events, partly through the social changes triggered by the rise of industrial mass societies, and partly through the undermining of traditional belief systems. Nonetheless, it can be argued that much misery throughout human history could have been avoided by the application of reason, and that the disasters of the 20th century resulted from a failure to be rational rather than a failure of rationality. The wise application of scientific understanding of the world in which we live is the only hope for the future of mankind.

The study of evolution has revealed our intimate connections with the other species that inhabit the Earth; if global catastrophe is to be avoided, these connections must be respected. The purpose of this book is to introduce the general reader to some of the most important basic findings, concepts, and procedures of evolutionary biology, as it has developed since the first publications of Darwin and Wallace on the subject, over 140 years ago. Evolution provides a set of unifying principles for the whole of biology; it also illuminates the relation of human beings to the universe and to each other. In addition, many aspects of evolution have practical importance; for instance, pressing medical problems are posed by the rapid evolution of resistance by bacteria to antibiotics and of HIV to antiviral drugs.



In this book, we shall first introduce the main causal processes of evolution (Chapter 2). Chapter 3 provides some of the basic biological background, and shows how the similarities between living creatures can be understood in terms of evolution. Chapter 4 describes the evidence for evolution derived from Earth history, and from the patterns of geographical distribution of living species. Chapter 5 is concerned with the evolution of adaptations by natural selection, and Chapter 6 with the evolution of new species and of differences between species. In Chapter 7, we discuss some seemingly difficult problems for the theory of evolution. Chapter 8 provides a brief summary.

# Chapter 2
# The processes of evolution

To understand life on Earth, we need to know how animals (including humans), plants, and microbes work, ultimately in terms of the molecular processes that underlie their functioning. This is the 'how' question of biology; an enormous amount of research during the last century has produced spectacular progress towards answering this question. This effort has shown that even the simplest organism capable of independent existence, a bacterial cell, is a machine of great complexity, with thousands of different protein molecules that act in a coordinated fashion to fulfil the functions necessary for the cell to survive, and to divide to produce two daughter cells (see Chapter 3). This complexity is even greater in higher organisms such as a fly or human being. These start life as a single cell, formed by the fusion of an egg and a sperm. There is then a delicately controlled series of cell divisions, accompanied by the differentiation of the resulting cells into many distinct types. The process of development eventually produces the adult organism, with its highly organized structure made up of different tissues and organs, and its capacity for elaborate behaviour. Our understanding of the molecular mechanisms that underlie this complexity of structure and function is rapidly expanding. Although there are still many unsolved problems, biologists are convinced that even the most complicated features of living creatures, such as human consciousness, reflect the operation of chemical and physical processes that are accessible to scientific analysis.

At all levels, from the structure and function of a single protein molecule, to the organization of the human brain, we see many instances of *adaptation*: the fit of structure to function that is also apparent in machines designed by people (see Chapter 5). We also see that different species have distinctive characteristics, often clearly reflecting adaptations to the environments in which they live. These observations raise the 'why' question of biology, which concerns the processes that have caused organisms to be the way they are. Before the rise of the idea of evolution, most biologists would have answered this question by appealing to a Creator. The term adaptation was introduced by 18th-century British theologians, who argued that the appearance of design in the features of living creatures proves the existence of a supernatural designer. While this argument was shown to be logically flawed by the philosopher David Hume in the middle of the 18th century, it retained its hold on people's minds as long as no credible alternative had been proposed.

Evolutionary ideas provide a set of natural processes that can explain the vast diversity of living species, and the characteristics that make them so well adapted to their environment, without any appeal to supernatural intervention. These explanations extend, of course, to the origin of the human species itself, and this has made biological evolution the most controversial of scientific subjects. If the issues are approached without prejudice, however, the evidence for evolution as an historical process can be seen to be as strong as that for other long-established scientific theories, such as the atomic nature of matter (see Chapters 3 and 4). We also have a set of well-verified ideas about the causes of evolution, although, as in every healthy science, there are unsolved problems, as well as new questions that arise as more is understood (see Chapter 7).

Biological evolution involves changes over time in the characteristics of populations of living organisms. The time-scale and magnitude of such changes vary enormously. Evolution can be studied during a human lifetime, when simple changes occur in a single character, such as the increase in the frequency of strains of bacteria resistant to penicillin within a few years of the widespread medical use of penicillin to control bacterial infections (as discussed in Chapter 5). At the other extreme, evolution involves events such as the emergence of a major new design of organisms, which may take millions of years and require changes in many different characteristics, as in the transition from reptiles to mammals (see Chapter 4). A key insight of the founders of evolutionary theory, Charles Darwin and Alfred Russel Wallace, was that changes at all levels are likely to involve the same types of processes. Major evolutionary changes largely reflect changes of the same type as more minor events, accumulated over longer time periods (see Chapters 6 and 7).

Evolutionary change ultimately relies on the appearance of new variant forms of organisms: *mutations*. These are caused by stable changes in the genetic material, transmitted from parent to offspring. Mutations affecting almost all conceivable characteristics of many different organisms have been studied in the laboratory by experimental geneticists, and medical geneticists have catalogued thousands of mutations in human populations. The effects of mutations on the observable characteristics of an organism vary greatly in their magnitude. Some have no detectable effect, and are known to exist simply because it is now possible to study the structure of the genetic material directly, as we will describe in Chapter 3. Others have relatively small effects on a simple trait, such as a change in eye colour from brown to blue, the acquisition of resistance to an antibiotic by a bacterium, or an alteration of the number of bristles on the side of a fruitfly. Some mutations have drastic effects on development, such as the mutation of the fruitfly *Drosophila melanogaster* that causes a leg to grow on the fly's head in place of its antenna. The appearance of any particular kind of new mutation is a very rare event, with a frequency of around one per hundred thousand individuals per generation or even less. An altered state of a character as a result of a mutation, such as antibiotic resistance, initially occurs in a single individual, and is usually restricted to a tiny fraction of a typical population for many generations. To result in evolutionary change, other processes must cause it to increase in frequency within the population.

*Natural selection* is the most important of these processes for evolutionary changes that involve the structure, functioning, and behaviour of organisms (see Chapter 5). In their papers of 1858, published in the *Journal of the Proceedings of the Linnaean Society*, Darwin and Wallace laid out their theory of evolution by natural selection with the following argument:

- Many more individuals of a species are born than can normally live to maturity and breed successfully, so that there is a *struggle for existence*.
- There is *individual variation* in innumerable characteristics of the population, some of which may affect an individual's ability to survive and reproduce. The successful parents of a given generation may therefore differ from the population as a whole.
- There is likely to be a *hereditary component* to much of this variation, so that the characteristics of the offspring of the successful parents will differ from the characteristics of the previous generation, in a similar way to their parents.



If this process continues from generation to generation, there will be a gradual transformation of the population, such that the frequencies of characteristics associated with greater survival ability or reproductive success increase over time. These altered characteristics originated by mutation, but mutations affecting a particular trait arise all the time regardless of whether or not they are favoured by selection. Indeed, most mutations either have no effects on the organism, or reduce its ability to survive or reproduce.

It is the process of increase in frequency of variants that improve survival or reproductive success that explains the evolution of adaptive characteristics, since better performance of the individual's body or behaviour will generally contribute to greater survival or reproductive success. Such a process of change will be especially likely if a population is exposed to a changed environment, where a somewhat different set of characteristics is favoured from those already established by selection. As Darwin wrote in 1858:

> But let the external conditions of a country alter . . . Now, can it be doubted, from the struggle each individual has to obtain subsistence, that any minute variation in structure, habits or instincts, adapting that individual better to the new conditions, would tell upon its vigour and health? In the struggle it would have a better *chance* of surviving; and those of its offspring that inherited the variation, be it ever so slight, would also have a better *chance*. Yearly more are bred than can survive; the smallest grain in the balance, in the long run, must tell on which death shall fall, and which shall survive. Let this work of selection on the one hand, and death on the other, go on for a thousand generations, who will pretend to affirm that it would produce no effect . . .



There is, however, another important mechanism of evolutionary change, which explains how species can also come to differ with respect to traits with little or no influence on the survival or reproductive success of their possessors, and which are therefore not subject to natural selection. As we shall see in Chapter 6, this is especially likely to be true of the large category of changes in the genetic material which have little or no effect on the organism's structure or functioning. If there is *selectively neutral* variability, so that on average there are no differences in survival or fertility among different individuals, it is still possible for the offspring generation to differ slightly from the parental generation. This is because, in the absence of selection, the genes in the population of offspring are a random sample of the genes present in the parental population. Real populations are finite in size, and so the constitution of the offspring population will by chance differ somewhat from that of the parents' generation, just as we do not expect exactly five heads and five tails when we toss a coin ten times.

This process of random change is called *genetic drift*. Even the biggest biological populations, such as those of bacteria, are finite, so that genetic drift will always operate.

The combined effects of mutation, natural selection and the random process of genetic drift cause changes in the composition of a population. Over a sufficiently long period of time, these cumulative effects alter the population's genetic make-up, and can thus greatly change the species' characteristics from those of its ancestors.

We referred earlier to the diversity of life, reflected in the large number of different species alive today. (A very much larger number have existed over the past history of life, owing to the fact that the ultimate fate of nearly all species is extinction, as described in Chapter 4.) The problem of how new species evolve is clearly a crucial one, and is dealt with in Chapter 6. The term 'species' is hard to define, and it is sometimes difficult to draw a clear line between populations that are members of the same species, and populations that belong to separate species. In thinking about evolution, it makes sense to consider two populations of sexually reproducing organisms as different species if they cannot interbreed with each other, so that their evolutionary fates are totally independent. Thus, human populations living in different parts of the world are unequivocally members of the same species, since there are no barriers to interbreeding if migrant individuals arrive from another place. Such migration tends to prevent the genetic makeup of different populations of the same species from diverging very much. In contrast, chimpanzees and humans are clearly separate species, since humans and chimpanzees living in the same area cannot interbreed. As we shall describe later on, humans also differ much more from chimpanzees in the make-up of their genetic material than they do from each other. The formation of a new species must involve the evolution of barriers to interbreeding between related populations. Once such barriers form, the populations can diverge under mutation, selection, and genetic drift. This process of divergence ultimately leads to the diversity of life. If we understand how barriers to interbreeding evolve, and how populations subsequently diverge, we will understand the origin of species.

An enormous amount of biological data falls into place in the light of these ideas about evolution, which have been put on a firm basis by the development of mathematical theories which can be modelled in detail, just as astronomers and physicists model the behaviour of stars, planets, molecules, and atoms in order to understand them more completely, and to devise detailed tests of their theories. Before describing the mechanisms of evolution in more detail (but omitting the mathematics), the next two chapters will show how many kinds of biological observations make sense in terms of evolution, in contrast with special creation and its appeal to *ad hoc* explanations.

# Chapter 3
# The evidence for evolution: similarities and differences between organisms

The theory of evolution accounts for the diversity of life, with all the well-known differences between different species of animals, plants, and microbes, but it also explains their fundamental similarities. These are often evident at the superficial level of externally visible characters, but extend to the finest details of microscopic structure and biochemical function. We will discuss the diversity of life later in this book (in Chapter 6), and describe how the theory of evolution can account for new forms appearing from ancestral ones, but here we focus on the unity of living species. In addition, we will introduce many basic biological facts on which later chapters build.

## Similarities between different groups of species

Similarities between even widely disparate types of organism exist at every level, from familiar, externally visible resemblances, to profound resemblances in life-cycles and the structure of the genetic material. They are plainly detectable even between creatures as different as ourselves and bacteria. These similarities have a natural and straightforward explanation in the idea that organisms are related through an evolutionary process of descent from common ancestors. We ourselves have obvious similarities to apes, as illustrated in Figure 1A, including similarities in internal characters such as our brain structure and organization. There are

A

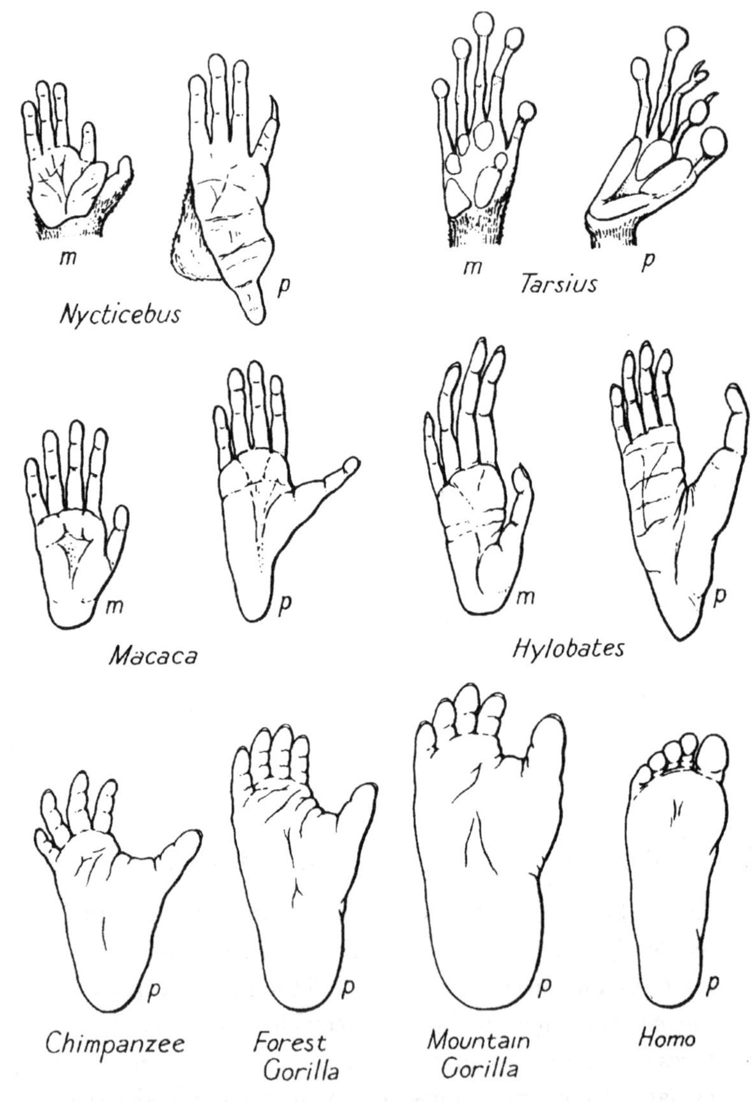


**1. A. Hands (*m*) and feet (*p*) of several primate species, showing the similarities between different species, with differences related to the animals' way of life, such as the opposable digits of climbing species (*Hylobates* is a gibbon, *Macaca* is a Rhesus monkey, *Nycticebus* and *Tarsius* are primitive arboreal primates). B. Skeletons of a bat and a bird, showing their similarities and differences.**

**B**

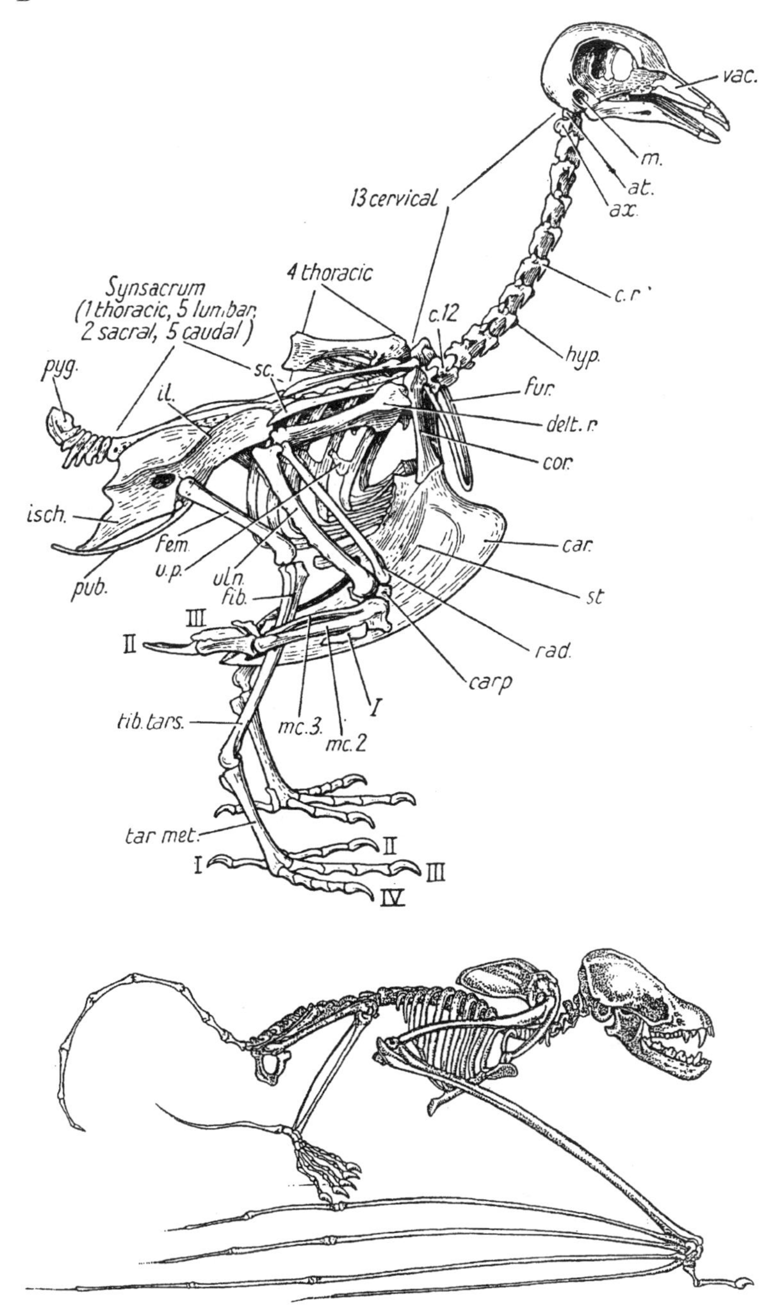
vac.
m.
at.
ax.
13 cervical
c.r
4 thoracic
c.12
hyp.
Synsacrum
(1 thoracic, 5 lumbar,
2 sacral, 5 caudal)
pyg.
sc.
il.
fur.
delt. r.
cor.
isch.
fem.
u.p.
car.
pub.
uln.
fib.
st
III
II
rad.
carp
I
tib. tars.
mc.3.
mc.2
tar met.
II
I
III
IV

lesser similarities to monkeys, and even smaller, but still extremely clear, similarities to other mammals, despite all our differences. Mammals have many similarities to other vertebrates, including the basic features of their skeletons, and their digestive, circulatory, and nervous systems. Even more amazing are similarities with creatures such as insects, for example in their segmented body plans, their common need for sleep, the control of their daily rhythms of sleep and waking, and fundamental similarities in how the nerves work in many different kinds of animals, among other features.

Systems of biological classification have long been based on easily visible structural characteristics. For example, even before the scientific study of biology, insects were treated as a group of similar creatures, clearly distinguishable from other groups of invertebrates, such as molluscs, by their possession of a segmented body, six pairs of jointed legs, a tough external protective covering, and so on. Many of these traits are shared with other types of animal such as crabs and spiders, except that the numbers of legs may differ (eight, in the case of spiders). These different species are all grouped into one larger division, the arthropods. The arthropods include the insects, and among these flies form one group, characterized by the fact that they all have only one pair of wings, as well as several other shared characters. Butterflies and moths form another insect group, whose members all have fine scales on their two pairs of wings. Among flies we distinguish the houseflies and their relatives from other groups by shared characters, and among these we name individual *species*, such as the common housefly *Musca domestica*. Species are essentially groups of similar individuals capable of interbreeding with each other. Similar species are grouped into the same *genus*, again united by a set of characters not shared with other genera. Biologists identify each recognizable species by two names, the genus name followed by the name of the species itself, for example *Homo sapiens*; these names are conventionally written in italics.

The observation that organisms can be classified hierarchically into groups, which successively share more and more traits that are lacking in other groups, was an important advance in biology. The classification of organisms into species, and the naming system for species, were developed long before Darwin. Before biologists could begin thinking about the evolution of species, it was clearly important to have the concept of species as distinct entities. The simplest and most natural way to account for the hierarchical pattern of similarities is that living organisms evolved over time, starting from ancestral forms that diversified to produce the groups alive today, as well as innumerable extinct organisms (see Chapter 4). As we shall discuss in Chapter 6, it is now possible to discern this inferred pattern of genealogical relationships among groups of organisms by directly studying the information in their genetic material.

Another set of facts that strongly supports the theory of evolution is provided by modifications of the same structure in different species. For instance, the bones of bats' and birds' wings indicate clearly that they are modified forelimbs, even though they look very different from the forelimbs of other vertebrates (Figure 1B). Similarly, although the flippers of whales look much like fish fins, and are clearly also well adapted for swimming, their internal structure is like the feet of other mammals, except for an increased number of digits. This makes sense, given all the other evidence that whales are modified mammals (for instance, they breathe with lungs and suckle their young). Fossil evidence shows that the two pairs of limbs of land vertebrates are derived from the two pairs of fins of the lobe-finned fishes (of which coelacanths are the most famous living representatives, see Chapter 4). Indeed, the earliest land vertebrate fossils had more than five digits on their limbs, just like fishes and whales. Another example is provided by the three small bones in mammals' ears, which transmit sound from the outside to the organ that transforms sound into nerve signals. These tiny bones develop from rudiments in the embryonic jaw and skull, and in reptiles they enlarge during development to make parts of the head and jaw skeleton. Fossil intermediates that connect reptiles with mammals show successive modifications of these bones in the adults, finally evolving into the ear bones. These examples are just a few of many known cases in which the same basic structure was considerably modified during the course of evolution by the demands imposed by different functions.

## Embryonic development and vestigial organs

Embryonic development provides many other striking examples of similarities between different groups of organisms, clearly suggesting descent from common ancestors. The embryonic forms of different species are often extremely similar, even when the adults are very different. For example, at one stage in mammalian development, gill slits appear that resemble those of fish embryos (Figure 2). This makes perfect sense if we are descended from fish-like ancestors, but is otherwise inexplicable. Since it is the adult structures that adapt the organism to its environment, they are very likely to be modified by selection. Probably the developing blood vessels require the presence of gill slits to guide them to form in the correct places, so that these structures are retained, even in animals that never have functional gills. Development can evolve, however. In many other details, mammals develop very differently from fish, so that other embryonic structures, with less profound importance in development, have been lost, and new ones have been gained.

Similarities are not confined to embryonic stages. *Vestigial organs* have also long been recognized as remnants of structures that were functional in the ancestors of present-day organisms. Their evolution is very interesting, because such cases tell us that evolution does not always create and improve structures, but sometimes reduces them. The human appendix, which is a greatly reduced version of a part of the digestive tract that is quite large in

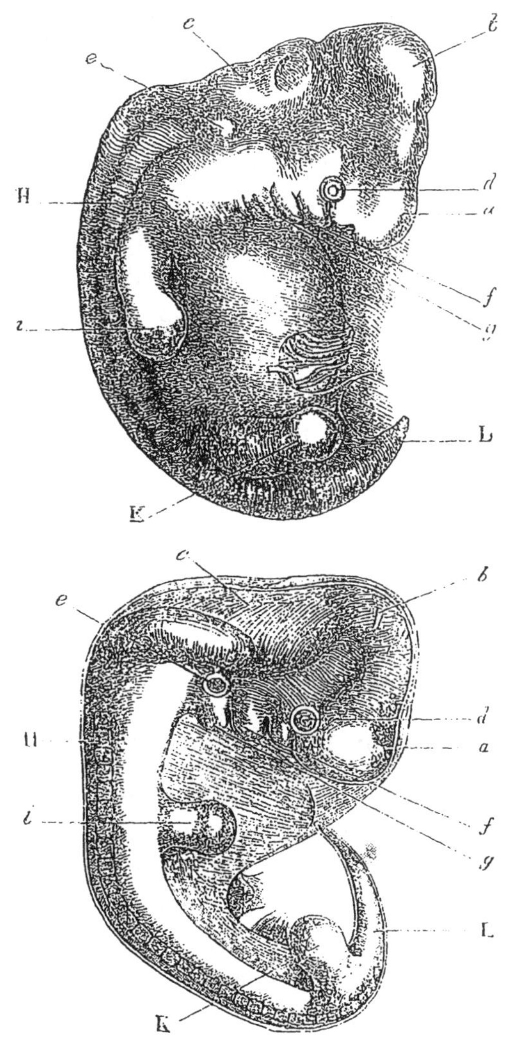


Fig. 1. Upper figure human embryo, from Ecker. Lower figure that of a dog, from Bischoff.

*a*. Fore-brain, cerebral hemispheres, &c.
*b*. Mid-brain, corpora quadrigemina.
*c*. Hind-brain, cerebellum, medulla oblongata.
*d*. Eye.
*e*. Ear.
*f*. First visceral arch.
*g*. Second visceral arch.
H. Vertebral columns and muscles in process of development.
*i*. Anterior } extremities.
K. Posterior } extremities.
L. Tail or os coccyx.

**2. Human and dog embryos, illustrating their great similarity at this stage of development. The gill slits, labelled visceral arches (f and g) in the figure, are plainly visible. From Darwin's *The Descent of Man and Selection in Relation to Sex* (1871).**

orang-utans, is a classic example. The vestigial limbs of legless animals are also well known. Fossils of primitive snakes have been found with almost complete hindlimbs, indicating that snakes evolved from lizard-like ancestors with legs. The body of a present-day snake consists of an elongated thorax (chest), with a large number of vertebrae (more than 300 in pythons). In the python, the change from the body to tail is marked by vertebrae with no ribs, and at this point rudimentary hindlimbs are found. There is a pelvic girdle and a pair of truncated thigh bones whose development follows the normal course for other vertebrates, with expression of the same genes that normally control limb development. A graft of python hindlimb tissue can even promote the formation of an extra digit in chick wings, showing that parts of the hindlimb developmental system still exist in pythons. More advanced types of snakes, however, are completely limbless.



## Similarities in cells and cellular functions

The similarities between different organisms are not confined to visible characteristics. They are profound and extend to the smallest microscopic scale and to the most fundamental aspects of life. A basic feature of all animal, plant, and fungal life is that their tissues are made up of essentially similar units, the *cells*. Cells are the basis of the bodies of all organisms other than viruses, from unicellular yeasts and bacteria, to multicellular bodies with highly differentiated tissues like those of mammals. In the *eukaryotes* (all cellular non-bacterial life) the cells are organized into the *cytoplasm* and the *nucleus* within it that contains the genetic material (Figure 3). The cytoplasm is not just a liquid inside the cell membrane with the nucleus floating in it; it contains a complex set of tiny pieces of machinery that includes many subcellular structures. Two of the most important of these cellular *organelles* are the mitochondria that generate cells' energy, and the chloroplasts in which photosynthesis in green plants' cells occurs. It is now known that both these are descended from bacteria that colonized cells and became integrated into them as essential components. Bacteria are also cells (Figure 3), but simpler ones with no nucleus or organelles; they and similar organisms are called *prokaryotes*. The only non-cellular forms of life, the viruses, are parasites that reproduce inside the cells of other organisms, and consist simply of a protein coat surrounding the genetic material.

Cells are ultra-miniaturized and highly complex factories which make the chemicals that organisms need, generate energy from food sources, and produce bodily structures such as the bones of animals. Most of the 'machines' and many of the structures in these factories are *proteins*. Some proteins are *enzymes* that take a chemical and carry out a procedure on it, for example snipping a chemical compound into two components, like chemical scissors. The enzymes used in biological detergents snip up proteins (such as blood and sweat proteins) into small pieces that can be washed out of dirty clothes; similar enzymes in our gut break molecules in food into smaller pieces that can be taken up by cells. Other proteins in living organisms have storage or transport functions. The haemoglobin in red blood cells carries oxygen, and in the liver a protein called ferritin binds and stores iron. There are also structural proteins, such as the keratin that forms skin, hair, and fingernails. In addition, cells make proteins that communicate information to other cells and to other organs. Hormones are familiar communication proteins, which circulate in the blood and control many bodily functions. Other proteins are located on cell surfaces and are involved in communication with other cells. These interactions include signalling to control cell behaviour during development, communication between eggs and sperm in fertilization, and parasite recognition by the immune system.

Like any factory, cells are subject to complex controls. They respond to information from outside (by means of proteins that span the cell membrane, like keyholes which fit molecules from the outside

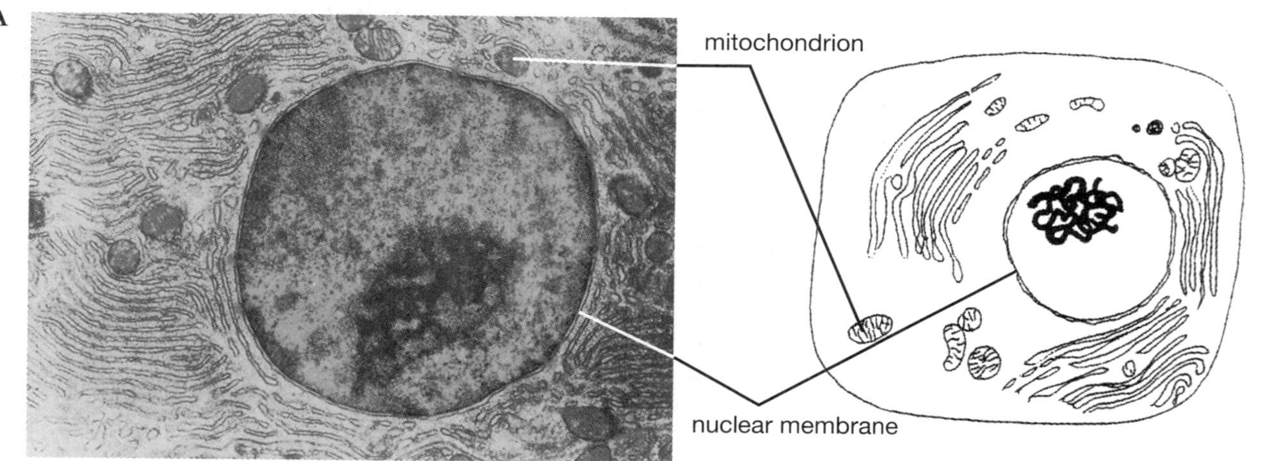


**3. Prokaryote and eukaryote cells.**

**A. Electron micrograph and drawing of a portion of a cell from the mammalian pancreas, showing the nucleus containing the chromosomes inside the nuclear membrane, the region outside the nucleus containing many mitochondria (these organelles also have membranes enclosing them), and membrane-like structures that are involved in protein synthesis and export, as well as in importing substances into the cell. A mitochondrion is somewhat smaller than a bacterial cell.**

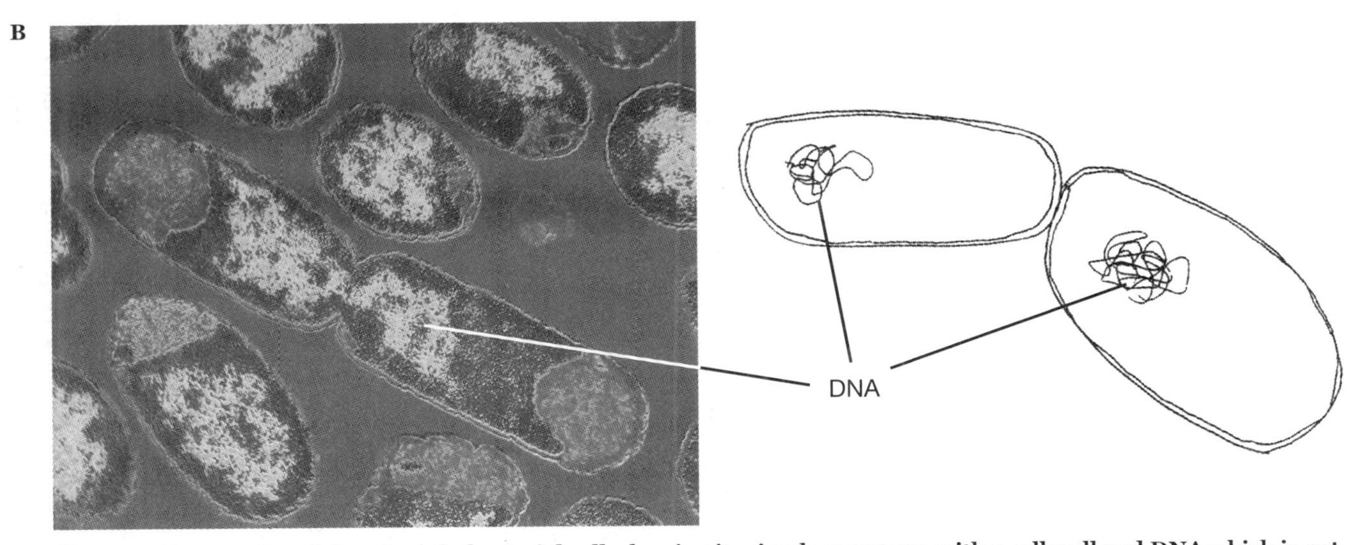


**B. Electron micrograph and drawing of a bacterial cell, showing its simple structure, with a cell wall and DNA which is not enclosed in a nucleus.**

world – see Figure 4). Sensory receptor proteins, such as olfactory receptors and light receptors, are used in communication between cells and their environment. Chemical and light signals from the outside world are transformed into electrical impulses that travel along the nerves to the brain. All animals that have been studied use largely similar proteins in chemical and light perception. To illustrate the similarities that have been discovered in cells of different organisms, a myosin (motor) protein, similar to proteins in muscle cells, is involved in signalling in flies' eyes and in the ears of humans; one form of deafness is caused by mutations in the gene for this protein.

Biochemists have catalogued the enzymes in living organisms into many different kinds, and every known enzyme (many thousands in a complex animal like ourselves) has a number in an international numbering system. Because so many enzymes are found in cells of a very wide range of organisms, this system categorizes enzymes by the jobs they perform, not the organism they come from. Some, such as digestive enzymes, snip molecules into pieces, others combine molecules together, while others oxidize chemicals (combine them with oxygen), and so on.



The means by which energy is generated by cells from food sources is largely the same for all kinds of cells. In this process, there is an energy source (sugars or fats, in the case of our cells, but other compounds, such as hydrogen sulphide, for some bacteria). A cell takes the initial compound through a series of chemical steps, some of which release energy. Such a *metabolic pathway* is organized like an assembly line, with a succession of sub-processes. Each sub-process is carried out by its own protein 'machine'; these are the enzymes for the different steps in the pathway. The same pathways operate in a wide range of organisms, and modern biology textbooks show the important metabolic pathways without needing to specify the organism. For example, when lizards tire after running, this is caused by the build-up of the chemical lactic acid, just as in our muscles. Cells have pathways to make chemicals of many different kinds, as well as to generate energy from foods. For example, some of our cells make hairs, some make bone, some make pigments, others produce hormones, and so on. The metabolic pathway by which the skin pigment melanin is made (Figure 4) is the same in ourselves, in other mammals, in butterflies with black wing pigments, and even in fungi (for instance in black spores), and many of the enzymes involved in this pathway are also used by plants in making lignin, the main chemical constituent of wood. The fundamental similarity of the basic features of metabolic pathways, from bacteria to mammals, is once again readily understandable in terms of evolution.

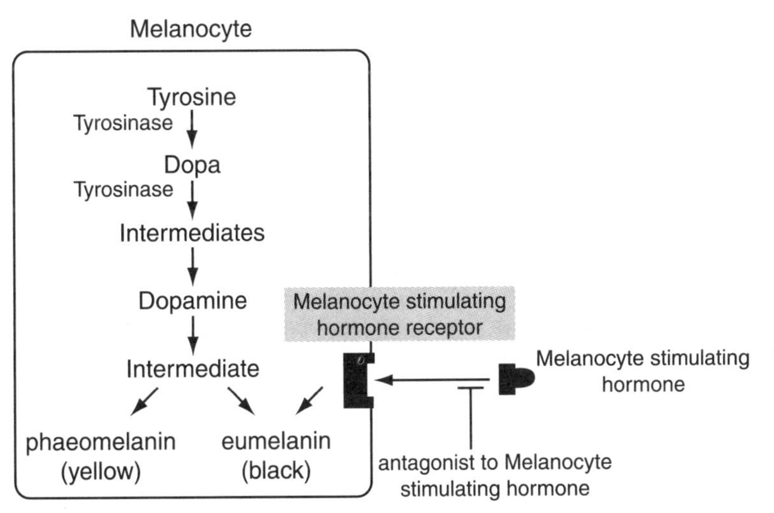


**4. Biosynthetic pathways by which melanin and a yellow pigment are synthesized in mammalian melanocyte cells from their amino acid precursor, tyrosine. Each step in the pathway is catalysed by a different enzyme. Absence of active tyrosinase enzyme results in albino animals. The melanocyte-stimulating hormone receptor determines the relative amounts of black and yellow pigments. Absence of the antagonist to the hormone leads to black pigment synthesis, but presence of the antagonist sets the receptor to 'off', leading to yellow pigment formation. This is how the yellow versus black parts of tabby cat and brown mouse hairs come to be formed. Mutations that make the antagonist non-functional cause darker coloration; however, black animals are not usually the result of this, but simply have the receptor set to 'on' regardless of the hormone level.**

Each of the different proteins for these cell and body functions is specified by one of the organism's genes, as we will explain more fully later in this chapter. The functioning of each biochemical pathway depends on its enzymes. If any enzyme in a pathway fails to work, the end-product will not be produced, just as a failure in an assembly-line process stops output of the product. For instance, albino mutations result from lack of an enzyme necessary for production of the pigment melanin (Figure 4). Stopping a step in a pathway is a useful means to control the output of the cell machinery, so cells contain inhibitors to carry out such control functions, as in the control of melanin production. As another example, the protein that forms blood clots is present in tissues, but in soluble form, and a clot will develop only when a piece is cut off this precursor molecule. The enzyme that cuts this protein is also present, but is normally inactive; when blood vessels are damaged, factors are released that alter the clotting enzyme, so that it immediately becomes active, leading to clotting of the protein.

**5. A. The three-dimensional structure of the protein myoglobin (a muscle protein similar to the red blood cell protein haemoglobin), showing the individual amino acids in the protein chain, numbered from 1 to 150, and the iron-containing haem molecule that the protein holds. The haem binds oxygen or carbon dioxide, and the protein's function is to carry these gas molecules.**

**B. The structure of DNA, the molecule that carries the genetic material in most organisms. It consists of two complementary strands, wound around each other in a helix. The backbone of each strand is formed of molecules of the sugar deoxyribose (S), linked to each other through phosphate molecules (P). Each sugar is connected to a type of molecule called a nucleotide; these form the 'letters' of the genetic alphabet. There are four types of nucleotide: adenine (A), guanine (G), cytosine (C), and thymine (T). A given nucleotide from one strand is paired with a complementary nucleotide from the other, as indicated by the double lines. The rule for this pairing is that A binds to T and G binds to C. When DNA replicates during cell division, the two strands unwind, and a complementary daughter strand is synthesized from each parental strand according to this pairing rule. In this way, a place where A and T bind to each other in the parental molecule produces a place with A and T in each of the daughter molecules.**

A

B

Proteins are very large molecules made up of strings of dozens to a few hundreds of *amino acid* subunits, each joined to a neighbouring amino acid, forming a chain (Figure 5A). Each amino acid is a quite complex molecule, with individual chemical properties and sizes. Twenty different amino acids are used in the proteins of living organisms; a particular protein, such as the haemoglobin in our red blood cells, has a characteristic set of amino acids in a particular order. Given the correct sequence of amino acids, the protein chain folds up into the shape of the working protein. The complex three-dimensional structure of a protein is completely determined by the sequence of amino acids in its constituent chain or chains; in turn, this sequence is completely determined by the sequence of chemical units of the *DNA* (Figure 5B) of the gene that produces the protein, as we will soon explain.

Studies of the three-dimensional structures of the same enzyme or protein in widely different species show that these are often extremely similar across huge evolutionary distances, such as between bacteria and mammals, even if the sequence of amino acids has changed greatly. An example is the myosin protein that we have already mentioned, which is involved in signalling in flies' eyes and in mammalian ears. Such fundamental similarities mean that, astonishingly, it is often possible to correct a metabolic defect in yeast cells by introducing a plant or animal gene with the same function. Yeast cells with a mutation causing a defect in ammonium uptake have been 'cured' by expressing a human gene in their cells (the gene for the Rhesus blood-group protein, RhGA, which was suspected to have the relevant function). The natural (non-mutant) yeast version of this protein has many amino acid differences from the human RhGA one, yet in this experiment the human protein can function in yeast cells lacking their own normal version. The result of this experiment also tells us that a protein with an altered amino acid sequence can sometimes work quite well.

## The basis of heredity is common to all organisms

The physical basis of inheritance is fundamentally similar in all eukaryote organisms (animals, plants, and fungi). Our understanding of the mechanism of inheritance, that is the control of individuals' many different characteristics by physical entities that we now call *genes*, first came from work by Gregor Mendel on

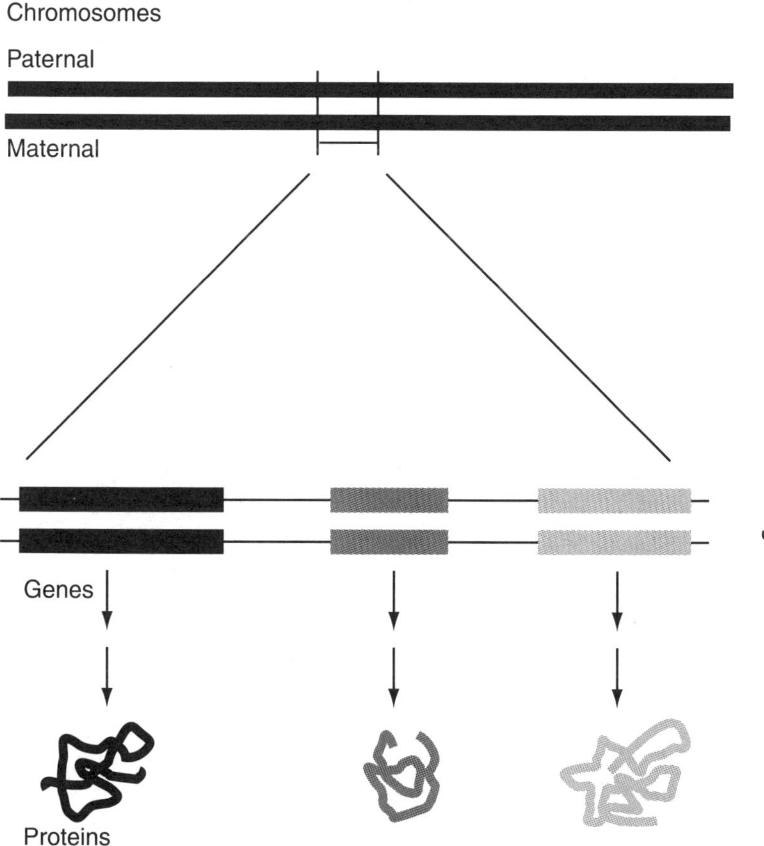


**6. Diagram of one pair of chromosomes, with a schematic drawing of a small region magnified to show three genes that are located in this chromosome region, and the non-coding DNA in between them. The three different genes are shown as different shades of grey, to indicate that each gene encodes a different protein. In a real cell, only some of the proteins would be produced, while other genes would be turned off so that their proteins would not be formed.**

garden peas, but the same rules of inheritance apply to other plants and to animals, including humans. The genes that control the production of metabolic enzymes and other proteins (and thus determine individuals' characteristics) are stretches of DNA carried in the *chromosomes* of each cell (Figures 6 and 7). The discovery that the chromosomes carry the organism's genes in a linear arrangement was first made in the fruitfly, *Drosophila melanogaster*, but it is equally true for our own genome. The order of genes on the chromosomes can be rearranged during evolution, but changes are infrequent, so that sets of the same genes in the same order can be found in the human genome and in the chromosomes of other mammals such as cats and dogs. A chromosome is essentially a single very long DNA molecule encoding hundreds or thousands of genes. The DNA of a



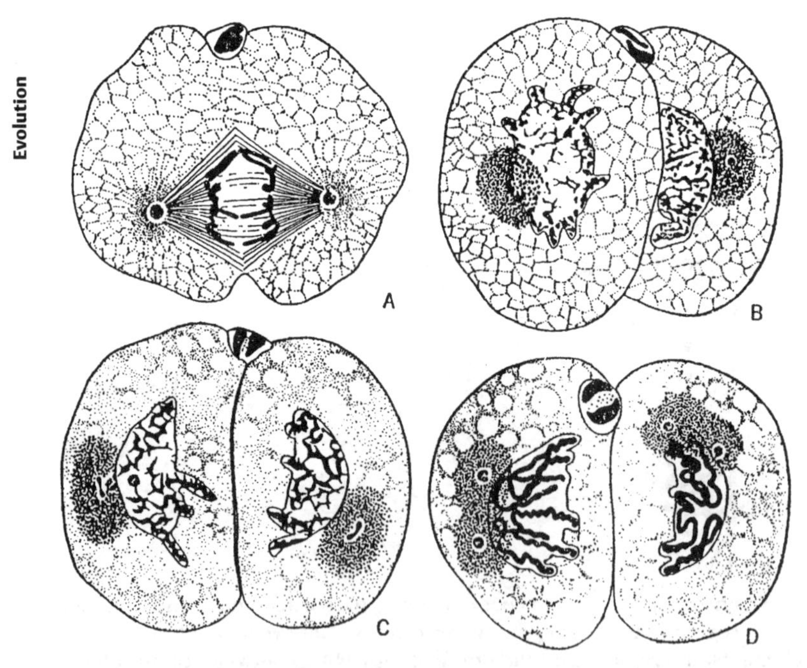


**7. A dividing cell of a nematode worm, showing the chromosomes no longer enclosed in the nuclear membrane (A), several stages in the division process (B, C), and finally the two daughter cells, each with a nucleus enclosed in a membrane (D).**

chromosome is combined with protein molecules that help to package it in neat coils inside the cell nucleus (resembling the devices used for keeping computer cables tidy).

In higher eukaryotes like ourselves, each cell contains one set of chromosomes derived from the mother through the egg nucleus, and another set derived from the father through the sperm nucleus (Figure 6). In humans, there are 23 different chromosomes in a single maternal or paternal set; in *Drosophila melanogaster*, which is used for much research in genetics, the chromosome number is five (one of which is tiny). The chromosomes carry the information needed to specify the amino acid sequences of an organism's proteins, together with the controlling DNA sequences that determine which proteins will be produced by the organism's cells.

What is a gene, and how does it determine the structure of a protein? A gene is a sequence of the four chemical 'letters' (Figure 5) of the *genetic code*, in which sets of three adjacent letters (*triplets*) correspond to each amino acid in the protein for which the gene is responsible (Figure 8). The gene sequence is 'translated' into the sequence of a protein chain; there are also triplets marking the end of the amino acid chain. A change in the sequence of a gene causes a mutation. Most such changes will lead to a different amino acid being placed in a protein when it is being made (but, because there are 64 possible triplets of DNA letters, and only 20 amino acids used in proteins, some mutations do not change the protein sequence). Across the entire range of living organisms, the genetic code differs only very slightly, strongly suggesting that all life on Earth may have a common ancestor. The genetic code was first studied in bacteria and viruses, but was soon checked and found to be the same in humans. Almost every possible mutation that this code can generate in the sequence of the human red blood cell protein haemoglobin has been observed, but mutations that are impossible with this particular code do not occur.

| Human | aac | cag | aca | gga | gcc | cgg | tgc | ctg | gag | gtg | tcc | atc | tct | gac | ggg | ctc | ttc | ctc | agc | ctg |
|---|---|---|---|---|---|---|---|---|---|---|---|---|---|---|---|---|---|---|---|---|
| Protein | Asn | Glu | Thr | Gly | Ala | Arg | Cys | Leu | Glu | Val | Ser | Ile | Ser | Asp | Gly | Leu | Phe | Leu | Ser | Leu |
| Human | aac | cag | aca | gga | gcc | cgg | tgc | ctg | gag | gtg | tcc | atc | tct | gac | ggg | ctc | ttc | ctc | agc | ctg |
| Chimpanzee | aac | cag | aca | gga | gcc | cgg | tgc | ctg | gag | gtg | tcc | atc | CcT | gac | ggg | ctc | ttc | ctc | agc | ctg |
| | | | | | | | | | | | | | Pro | | | | | | | |
| Dog | aac | cag | acC | ggG | Ccc | cgg | tgc | ctg | gag | gtg | tcc | att | CcC | Aac | ggg | ctG | ttc | ctc | agc | ctg |
| | | | * | * | Pro | | | | | | | | Pro | * | | * | | | | |
| Mouse | aac | cag | Tca | gAG | CcT | Tgg | tgc | ctg | TaT | gtg | tcc | atc | CcA | gaT | ggC | ctc | ttc | ctc | agc | ctA |
| | | | Ser | Glu | Pro | Trp | | | Tyr | | | | Pro | * | * | | | | | * |
| Pig | aac | cag | acG | ggC | Ccc | cAg | tgc | ctg | gag | gtg | tcc | atT | CCC | gac | ggg | ctc | ttc | ctc | agc | ctg |
| | | | | | Pro | Gln | | | | | | * | Pro | | | | | | | |

| Human | ggg | ctg | gtg | agc | ttg | gtg | gag | aac | gcg | ctg | gtg | gtg | gcc | acc | atc | gcc | aag | aac | cgg | aac |
|---|---|---|---|---|---|---|---|---|---|---|---|---|---|---|---|---|---|---|---|---|
| Protein | Leu | Gly | Leu | Val | Ser | Leu | Val | Glu | Asn | Ala | Leu | Val | Val | Ala | Thr | Ile | Ala | Lys | Asn | Arg |
| Human | ggg | ctg | gtg | agc | ttg | gtg | gag | aac | gCg | ctg | gtg | gtg | gcc | acc | atc | gcc | aag | aac | cgg | aac |
| Chimpanzee | ggg | ctg | gtg | agc | ttg | gtg | gag | aac | ATg | ctg | gtg | gtg | gcc | acc | atc | gcc | aag | aac | cgg | aac |
| | | | | | | | | | Met | | | | | | | | | | | |
| Dog | ggg | ctg | gtg | agc | Gtt | gtg | gaa | aaT | gTg | ctg | gtg | gtg | gcc | Gcc | atT | gcc | aag | aac | cgC | aa |
| | | | | | * | | | * | Val | | | | | * | * | | | | * | |
| Mouse | ggg | ctg | gtg | agt | Ctg | gtg | gag | aaT | gTg | ctg | gtT | gtg | ATA | Gcc | atc | Acc | aaa | aac | cgC | aac |
| | | | | * | * | | | * | Val | | * | | * | * | | | | | * | |
| Pig | ggg | ctg | gtg | agC | ctC | gtg | gag | aac | gTg | ctg | gtg | gtg | gcc | Gcc | atc | gcc | aag | aac | cgC | aac |
| | | | | | | | | | Val | | | | | * | | | | | * | |

**8. DNA and protein sequences of a part of the gene for the melanocyte-stimulating hormone receptor shown in Figure 4, in humans and several other mammals. The figure shows only 40 amino acids out of the total of 951 in the protein. The human DNA sequences are shown at the top, with spaces between the sets of three DNA letters, and the protein sequence is in the grey bars below this (using a three-letter code for the different amino acids). The other species are shown below. Where the DNA sequences differ from the human one, the letter is printed as a capital letter. Triplets that include a difference from the human sequence, but code for the same amino acid as in humans, are asterisked, while triplets that encode differences from the human protein sequence are highlighted. Many red-haired people have an amino acid variant in triplet 151.**

In order to produce its protein product, the DNA sequence of a gene is first copied into a 'message' made of a related molecule, *RNA*, whose sequence of 'letters' is copied from that of the gene by a copying enzyme. The RNA message interacts with an elaborate piece of cellular machinery, made up of a conglomeration of proteins and other RNA molecules, to translate the message and produce the protein specified by the gene. This process is essentially the same in all cells, although in eukaryotes it occurs in the cytoplasm, and the message must first move from the nucleus to the cell regions where the translation machinery is sited. In between the genes on the chromosomes are stretches of DNA which do not code for proteins; some of this *non-coding* DNA has the important function of acting as sites for binding proteins that turn the production of the RNA messages of genes on or off as needed. For instance, the genes for haemoglobin are turned on in cells developing into red blood cells, but off in brain cells.

Despite the enormous differences in the modes of life of different organisms, ranging from unicellular organisms to bodies composed of billions of cells with highly differentiated tissues, eukaryote cells undergo similar cell division processes. Single-celled organisms such as an amoeba or yeast can reproduce simply by division into two daughter cells. A fertilized egg of a multicellular organism, produced by the fusion of an egg and a sperm, similarly divides into two daughter cells (Figure 7). Many further rounds of cell divisions then take place to produce the many cells and tissue types that form the body of the adult organism. In a mammal, there are over 300 different types of cell in the adult body. Each type has a characteristic structure and produces a specific array of proteins. The arrangement of these cells into tissues and organs during development requires elaborately controlled networks of interactions between the cells of the developing embryo. Genes are turned on and off to ensure that the right kind of cell is produced in the right place at the right time. In some well-studied organisms, such as *Drosophila melanogaster*, we now know a great deal about how these interactions result in the emergence of the intricate body plan of the fly from the apparently undifferentiated egg cell. Many signalling processes involved in development and differentiation of particular tissues, such as nerves, are found to be shared by all multicellular animals, while land plants use a rather different set, as might be expected from the fact that the fossil record shows that multicellular animals and plants have separate evolutionary origins (see Chapter 4).

When a cell divides, the DNA of the chromosomes is first replicated, so that there are two copies of each chromosome. Cell division is a process with tight controls to ensure that the newly copied DNA sequence undergoes 'proof-reading' for errors. Cells have enzymes that, using certain properties of the way DNA is replicated, can distinguish new DNA from the old 'template' DNA. This enables most errors in copying to be detected and corrected, ensuring that the template has been faithfully copied before the cell is allowed to proceed to the next step, division of the cell itself. The machinery of cell division ensures that each daughter cell receives a complete copy of the set of chromosomes that was present in the parent cell (Figure 7).

Most prokaryotes' genes (including those of many viruses) are also sequences of DNA which are organized only slightly differently from those carried in eukaryote chromosomes. Many bacteria have just one circular DNA molecule as their genetic material. Some viruses, however, such as those responsible for influenza and AIDS, have genes made of RNA. The proof-reading that occurs in DNA replication does not happen when RNA is copied, and so these viruses have extremely high mutation rates, and can evolve very rapidly within the host's body. As we will describe in Chapter 5, this means that it is difficult to develop vaccines against them.

Eukaryotes and prokaryotes differ greatly in their amounts of non-coding DNA. The bacterium *Escherichia coli* (a normally harmless species that lives in our intestines) has about 4,300 genes, and the stretches that code for protein sequences form about 86% of this species' DNA. In contrast, less than 2% of the DNA in the human genome codes for protein sequences. Other organisms lie between these extremes. The fruitfly, *Drosophila melanogaster*, has about 14,000 genes in about 120 million 'letters' of DNA, and about 20% of the DNA is made up of coding sequences. The number of different genes in the human genome is still not precisely known. The current best count comes from the sequencing of the complete genome. This allows geneticists to recognize sequences that are probably genes, based on what we know from genes that had previously been studied. It is a difficult task to find these sequences in the huge amount of DNA that makes up the genome of any species, particularly for our own genome, which has a very large DNA content (25 times as much as the fruitfly). The number of genes in humans is about 35,000, much smaller than had been guessed from the number of cell and tissue types with different functions. The number of proteins a human can make is probably considerably larger than this, because this method of counting cannot detect very small genes, or unconventional ones (for example, genes that lie within other genes, which exist in several organisms). It is not yet known how much of the non-coding DNA is important for the life of the organism. Although much of it is made up of viruses and other parasitic entities that live in chromosomes, some of it has important functions. As we have already mentioned, there are DNA sequences outside genes that can bind proteins controlling which genes in a cell are 'turned on'. The control of gene activity must be much more important in multicellular creatures than in bacteria.

In addition to the discovery that widely different organisms have DNA as their genetic material, modern biology has also uncovered profound similarities in the life-cycles of eukaryotes, despite their diversity, which ranges from unicellular fungi such as yeasts, to annual plants and animals, to long-lived (though not immortal) creatures like ourselves and many trees. Many, though not all, eukaryotes have a sexual stage in each generation, in which the maternal and paternal genomes of the uniting egg and sperm (each made up of a set of some number $n$ of different chromosomes, characteristic of the species in question) combine to make an individual with $2n$ chromosomes. When an animal makes new eggs or sperm, the $n$ condition is restored by a special kind of cell division. Here, each pair of maternal and paternal chromosomes lines up, and (after exchanging material to form chromosomes that are patchworks partly of paternal and partly of maternal DNA) the chromosome pairs separate from each other in a similar way to the separation of newly replicated chromosomes in other cell divisions. At the end of the process, the number of chromosomes in each egg or sperm cell nucleus is therefore halved, but each egg or sperm has one complete set of the organism's genes. The double set will be restored on the union of egg and sperm nuclei at fertilization.



The basic features of sexual reproduction must have evolved long before the evolution of multicellular animals and plants, which are latecomers on the evolutionary scene. This is clear from the common features displayed in the reproduction of sexual unicellular and multicellular organisms, and the similar genes and proteins that have been discovered to be involved in the control of cell division and chromosome behaviour in groups as distant as yeast and mammals. In most single-celled eukaryotes, the $2n$ cell produced by fusion of a pair of cells, each with $n$ chromosomes, divides immediately to produce cells with $n$ chromosomes, just as described above for germ cell production in multicellular animals. In plants, the reduction of chromosome number from $2n$ to $n$ happens before egg and sperm formation, but the same kind of special cell division is again involved; in mosses, for instance, there is a prolonged life-cycle stage with chromosome number $n$ that forms the moss plant, on which the small $2n$ parasitic stage develops after eggs and sperm are made and fertilization has occurred.

The complications of such sexual processes are absent from some multicellular organisms. In such 'asexual' species, mothers produce daughters without a reduction of chromosome number from $2n$ during egg production. Nevertheless, all multicellular asexual organisms show clear signs of being descended from sexual ancestors. For example, common dandelions are asexual; their seeds form without the need for pollen to be brought to the flowers, as is required for most plants to reproduce. This is an advantage to a weedy species like the common dandelion, which speedily generates large numbers of seeds, as anyone who has a lawn can see for themselves. Other dandelion species reproduce by normal matings between individuals, and common dandelions are so closely related to these that they still make pollen that can fertilize the flowers of the sexual species.

## Mutations and their effects

Despite the proof-reading mechanisms that correct errors when DNA is copied during cell division, mistakes do occur, and these are the source of mutations. If a mutation results in a change in the amino acid sequence of a protein, the protein may malfunction; for example, it may not fold up correctly and so may be unable to do its job properly. If it is an enzyme, this can cause the metabolic pathway to which it belongs to run slowly, or not at all, as in the case of the albino mutations already mentioned. Mutations in structural or communication proteins may impair cell functions or the organism's development. Many diseases in humans are caused by such mutations. For instance, mutations in genes involved in controlling cell division increase the risk of cancer developing. As already mentioned, cells have exquisite control systems to ensure that they divide only when everything is in order (proof-reading for mutations must be complete, the cell must show no signs of infection or other damage, and so on). Mutations affecting these control systems can result in uncontrolled cell division, and malignant growth of the cell lineage. Luckily, it is unusual for both members of a pair of genes in a cell to be mutant, and one non-mutant member of the pair is often enough for correct cell functioning. A cell lineage also usually requires other adaptations to become a successful cancer, so malignancy is uncommon. (A blood supply is needed for tumours, and the cells' abnormal characteristics must evade detection by the body.) Nevertheless, understanding cell division and its control is a major part of cancer research. The process is so similar in cells of different eukaryote organisms that the 2001 Nobel prizes in medicine were given for research on cell division in yeast, which showed that a gene involved in the control system of yeast cells is mutated in some human familial cancers.

Mutations that give a predisposition to cancer are rare, as are most other disease-causing mutations. The most common genetic disorder in northern European human populations is cystic fibrosis, but even in this case the non-mutant sequence of the gene involved represents more than 98% of copies of the gene in the population. Mutations that cause failure of an important enzyme or protein may lower the survival or fertility of affected individuals. The gene sequence that leads to the non-functional enzyme will thus be under-represented in the next generation, and will eventually be eliminated from the population. A major role of natural selection is to keep the proteins and other enzymes of most individuals working well. We will revisit this idea in Chapter 5.

One important type of mutation leads to a protein not being produced in sufficient amounts by its gene. This could happen because of a problem in the normal control system for that gene, which either does not switch it on when it should do so, does not produce in the right quantities, or stops production of the protein before it is finished. Other mutations may not abolish an enzyme's production, but the enzyme may be faulty, just as a production line can be hindered or stopped if one of the necessary tools or machines is defective in some way. If one or more of the component amino acids are missing, the protein may not function correctly, and the same can happen if a different amino acid appears at a particular position in the chain, even if all the rest are correct. Mutations causing loss of function can contribute to evolution when selection no longer acts to eliminate them (see Chapters 2 and 6 for how selectively neutral mutations can spread). About 65% of human olfactory receptor genes are 'vestigial genes' that do not produce working receptor proteins, so we have many fewer olfactory functions than mice or dogs (not surprisingly, given the importance of smell in their daily lives and social interactions, compared with its minor role in ours).

There are also many differences between normal individuals in a species. For instance, individuals in human populations differ in their ability to taste or smell certain chemicals, or to break down some chemicals used as anaesthetics. People who lack an enzyme that breaks down an anaesthetic may suffer a bad reaction to it, but the lack of the enzyme would otherwise not matter. Similar differences in the ability to deal with other drugs, and sometimes foods, are an important aspect of variability in humans, and knowledge of these differences is necessary for modern medicine, in which strong drugs are often used.

Mutations in the enzyme glucose-6-phosphate dehydrogenase (an enzyme for an early step in the pathway by which cells derive energy from glucose) illustrate some of these kinds of differences. Individuals entirely missing this gene cannot survive (because the pathway in which it functions is vital in controlling the levels of toxic chemicals produced as a by-product of cellular energy generation). In human populations, there are at least 34 different normal variants of the protein that are not only compatible with healthy life, but are actually protective against malaria parasites. Each of these differs by one or a few amino acids from the protein's most common normal sequence. Several of these variants are widespread in Africa and the Mediterranean regions, and in some malarial populations variant individuals are frequent. However, some of the variants cause a form of anaemia when a type of bean is eaten, or when certain anti-malarial drugs are given. The well known ABO and other blood-groups are another example of normal variability within the human population; they are due to variation in the sequences of proteins that control details of the surfaces of red blood cells. Variation in the receptor protein for melanocyte-stimulating hormone, which is important in the production of the skin pigment melanin (see Figure 4), can cause hair colour differences; in many red-haired people, this protein has an altered amino acid sequence. As we shall discuss in Chapter 5, genetic variability is the essential raw material on which natural selection acts to produce evolutionary changes.

## Biological classification and DNA and protein sequences

A new and important set of data providing clear evidence that organisms are related to one another through evolution comes from the letters in their DNA, which can now be "read" by the chemical procedure of DNA sequencing. Systems of biological classification based on visible characteristics, which were developed over the past three centuries of study of plants and animals, are now supported by recent work comparing DNA and protein sequences among different species. Measuring the similarity of DNA sequences makes it possible to have an objective concept of relationship among species. We will describe this in more detail in Chapter 6. For the moment we need only understand that the DNA sequences of a given gene will be most similar for more closely related species, while those of more distantly related species are more different (Figure 8). The amount of difference increases roughly proportionally to the amount of time separating two sequences being compared. This property of molecular evolution allows evolutionary biologists to estimate times of events that cannot be studied in fossils, using a *molecular clock*. For instance, we have already mentioned changes in the order of an organism's genes on its chromosomes. A molecular clock can be used to estimate the rate of such chromosomal rearrangements. Consistent with the evolutionary viewpoint, species that we believe to be close relatives, such as humans and rhesus monkeys, have chromosomes that differ by fewer rearrangements than humans and New World primates such as the woolly monkey.

In the next chapter, we will explain the evidence for evolution based on fossil data, and from data on the geographical distribution of living species. These observations complement those described here, in showing that the theory of evolution provides a natural explanation for a wide range of biological phenomena.

# Chapter 4
# The evidence for evolution: patterns in time and space

> The history of man, therefore, is but a short ripple in the ocean of time.
>
> From *On the Interaction of the Natural Forces*,
> Hermann von Helmholtz, 1854

## The age of the Earth

It would have been impossible to realize that living organisms have originated by evolution, without the success of late 18th- and early 19th-century geologists in establishing that the present-day structure of the Earth is itself the product of long-continued physical processes. The methods involved are similar in principle to those used by historians and archaeologists. As the great French naturalist, the Comte de Buffon, wrote in 1774:

> Just as in civil history we consult warrants, study medallions, and decipher ancient inscriptions, in order to determine the epochs of human revolutions and fix the date of moral events, so in natural history one must dig through the archives of the world, extract ancient relics from the bowels of the earth, gather together their fragments and assemble again in a single body of proofs all those indications of the physical changes which can carry us back to the different Ages of Nature. This is the only way of fixing certain points in the immensity of space, and of placing milestones on the eternal path of time.

At the risk of some oversimplification, there were two key insights that led to the successes of early geology; the principle of *uniformitarianism*, and the invention of *stratigraphy* as a method of dating. Uniformitarianism is particularly associated with the late 18th-century Edinburgh geologist James Hutton, and was codified later by another Scottish scientist, Charles Lyell, in his *Principles of Geology* (1830). It is simply the application to the history of the structure of the Earth of the same principle used by astronomers in attempting to understand the constitution of distant planets and stars: the basic physical processes involved are assumed to be the same everywhere and at all times. Geological change over time reflects the operation of the laws of physics, which are themselves unchanging. For example, physical theory implies that the speed of rotation of the Earth must have decreased over millions of years because of frictional forces induced by the tides, which are caused by the gravitational forces of the Sun and Moon. The length of the day is now much longer than when the Earth was first formed, but the strength of the force of gravity has not changed.

There is, of course, no independent justification of this assumption of uniformity, any more than there is any logical justification for the assumption of the regularity of nature that underlies the most basic aspects of our daily life. Indeed, there is no distinction between these two assumptions, except the scale of time and space to which they apply. Their justification is that, first, uniformitarianism represents the simplest possible basis on which we can proceed to interpret events that are remote in time and space. Second, it has proved to be remarkably successful.

The uniformitarian assumption in geology implies that the present-day constitution of the Earth's surface reflects the cumulative action of processes of formation of new rocks by volcanic action and deposition of sediments in rivers, lakes, and seas, and the erosion of old rocks by the action of wind, water, and ice. The formation of *sedimentary* rocks like sandstone and limestone depends on the erosion of other rocks. In contrast, the formation of mountains by volcanic action and uplift of land by earthquakes must precede their degradation by erosion. These processes can be observed in action in the present day; anyone who has visited a mountainous region, especially at a time of year when freezing and thawing is happening, will have witnessed erosion of rocks, and the transport of the resulting debris downstream by streams and rivers. The deposition of sediments at the mouths of rivers is also easy to observe. Volcanic and earthquake activity are confined to certain regions of the globe, especially the edges of continents and middles of oceans, for reasons which are now well understood, but there are numerous recorded instances of the formation of new oceanic islands by volcanic action, and of the uplift of land by earthquakes. In *The Voyage of the Beagle*, Darwin described the effects of the Chilean earthquake of February 1835 in the following terms:



> The most remarkable effect of this earthquake was the permanent elevation of the land; it would probably be more correct to speak of it as the cause. There can be no doubt that the land around the Bay of Concepcion was upraised two or three feet . . . At the island of S. Maria (about thirty miles distant) the elevation was greater; on one part Captain Fitzroy found beds of putrid mussel-shells *still adhering to the rocks*, ten feet above high water mark . . . The elevation of this province is particularly interesting from its having been the theatre of several other violent earthquakes, and from the vast number of sea-shells scattered over the land, up to a height of certainly 600, and I believe, 1000 feet.

Geology has been outstandingly successful in interpreting the structure of the Earth at or near its surface in terms of these processes, and in reconstructing the events that have led to the present-day appearance of many parts of the Earth. The order of these events can be established by the principle of stratigraphy. Information on the mineral composition and arrays of fossils found in different layers of rocks (*strata*) is used to characterize individual layers. The recognition that fossils represent the preserved remains of long-dead plants and animals, rather than artefacts of mineral formation, was essential for the success of stratigraphy. The types of fossils found in a given sedimentary rock layer provide evidence about the environment that prevailed when it was laid down; for example, it is usually possible to tell whether the organisms were marine, freshwater, or terrestrial. Fossils are, of course, absent from rocks such as granite or basalt that form by the solidification of molten material from below the Earth's crust.

During his travels throughout Britain to construct canals in the early 19th century, the English canal engineer William Smith recognized that similar successions of strata occur in different parts of Britain (which has an unusual variety of rocks of different ages for such a small area of land). Using the principle that older rocks must normally lie below younger ones, the comparison of the succession of strata in different localities enabled geologists to reconstruct sequences of strata that were laid down through immense periods of time. If rocks of type A are found below type B in one location, and B is found below C somewhere else, then one infers the sequence A-B-C, even if A and C are never found together in one place.



The systematic use of this method by 19th-century geologists allowed them to determine the major divisions of geological time (Figure 9). These yield a relative, not an absolute, chronology; absolute dates require methods for calibrating the rate of the processes involved, which is very difficult to do with any precision. The processes involved in landscape formation are very slow; erosion of even a few millimetres of rock takes many years, and the deposition of sediments is correspondingly slow. Similarly, uplift of land even in the most active mountain-building areas, such as the Andes, occurs at a rate of only a fraction of a metre a year on average. Given the existence of sedimentary rocks of the same formation that are several kilometres deep in many parts of the

| Era | Period | Epoch | years ago |
|---|---|---|---|
| | Quaternary | Holocene | 10,000 |
| | | Pleistocene | 2 million |
| Cenozoic | Tertiary | Pliocene | 7 million |
| | | Miocene | 26 million |
| | | Oligocene | 38 million |
| | | Eocene | 54 million |
| | | Palaeocene | 64 million |
| Mesozoic | Cretaceous | | 136 million |
| | Jurassic | | 190 million |
| | Triassic | | 225 million |
| Palaeozoic | Permian | | 280 million |
| | Carboniferous | | 345 million |
| | Devonian | | 410 million |
| | Silurian | | 440 million |
| | Ordovician | | 530 million |
| | Cambrian | | 570 million |

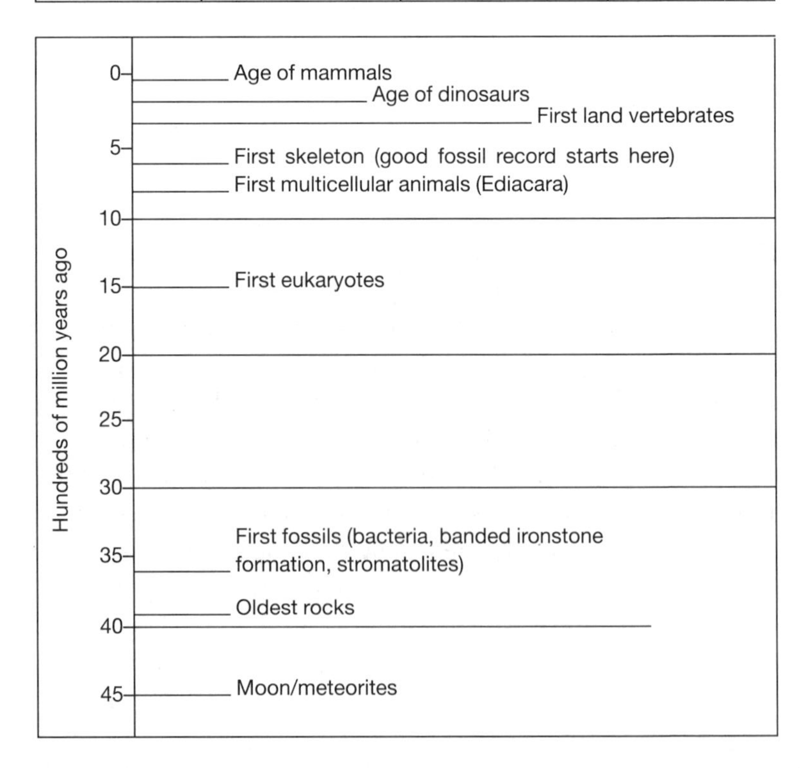

world, and the evidence that equally large deposits have been eroded, the necessity of a time-scale of at least many tens of millions of years for the existence of the Earth was quickly recognized, in conflict with Biblical chronology. Lyell suggested on this basis that the Tertiary period lasted about 80 million years, and that the Cambrian occurred 240 million years ago.

Such a long time-scale for the Earth was challenged by the eminent physicist Lord Kelvin, on the grounds that the rate of cooling of an originally molten Earth would make the Earth's core much cooler than it actually is, if the Earth had been formed more than approximately 100 million years ago. Kelvin's calculation was correct for the physics of his day. However, at the end of the 19th century, radioactive decay of unstable elements, such as uranium, into more stable derivatives was discovered. This process of decay is accompanied by the release of energy sufficient to slow the rate of cooling of the Earth to a value that agrees with current estimates of its age.

Radioactivity also provided new and reliable methods for establishing the ages of rock samples. The atoms of radioactive elements have a constant probability per year of decaying to a more stable daughter element, accompanied by the emission of radiation. When a rock is laid down, it can be assumed that the element in question is pure; hence, if the proportion of the daughter element in the sample is measured, the time since the formation of the rock can be estimated, knowing the rate of the decay process as established by experiments. Different elements are useful for dating rocks of different ages. Determinations of the ages of rocks belonging to the different periods of geology by this technique have given us the

**9. The major divisions of geological time. The upper part shows the named divisions from the Cambrian onwards, in which most fossils are found (this is less than one-eighth of the time since the formation of the Earth). The lower part shows the major events that have occurred during Earth history.**

dates accepted today. While the methods are constantly being refined, and the dates revised, the general time-scale that they indicate is very clear (Figure 9). It establishes an immense, almost incomprehensible, amount of time for biological evolution to occur.

## The fossil record

The fossil record is our only direct source of information on the history of life. To interpret it correctly, it is necessary to understand how fossils are formed, and how scientists study them. When a plant, animal, or microbe dies, the soft parts are almost certain to decay rapidly. Only in unusual environments, such as the arid atmosphere of a desert or the preservative chemicals of a piece of amber, are the microbes responsible for decay unable to break down the soft parts. Remarkable cases of preservation of soft parts, sometimes going back tens of millions of years in the case of insects trapped in amber, have been found, but these are the exception rather than the rule. Even skeletal structures, such as the tough chitin which covers the bodies of insects and spiders, or the bones and teeth of vertebrates, eventually decay. Their slower rate of disappearance offers, however, an opportunity for minerals to infiltrate them, and eventually replace the original material with a mineralized replica (occasionally this happens to soft parts as well). Alternatively, they may create a mould of their shape as minerals are deposited around them.

Fossilization is most likely to happen in aquatic environments, where the deposition of sediment and precipitation of minerals occur at the bottom of seas, lakes, and river estuaries. Remains that sink to the bottom can then turn into fossils, although the chance that this happens for a given individual is extremely small. The fossil record is therefore very biased: marine organisms living in shallow seas, where sediments are continuously formed, have the best fossil record, and flying creatures have the worst. In addition, the deposition of sediments may be interrupted, for example by a change in climate or by uplift of the seabed. For many types of creature, we have almost no fossil record; for others, the record is interrupted many times.

An excellent example of the problems caused by this incompleteness is provided by the coelacanth. This is a type of bony fish with lobed fins, related to the ancestors of the first land vertebrates. Coelacanths were abundant in the Devonian era (400 million years ago), but subsequently declined in number. The last fossil coelacanths are dated to about 65 million years ago, and the group was long thought to be extinct. In 1939 fishermen from the Comoro islands off the south-eastern coast of Africa caught a strange-looking fish, which turned out to be a coelacanth. Scientists have subsequently been able to study the habits of living coelacanths, and a new population has been discovered in Indonesia. The coelacanths must have existed continuously over a vast stretch of time, but left no fossil record because of their low abundance and the great depth at which they live.



The gaps in the fossil record mean that it is rare to have a long-continued series of remains showing the more or less continuous changes which are expected under the hypothesis of evolution. In most cases, new groups of animals or plants make their first appearance in the fossil record without any obvious links to earlier forms. The most famous example is the 'Cambrian explosion', which refers to the fact that most of the major groups of animals appear for the first time as fossils in the Cambrian period, between 550 and 500 million years ago (this will be discussed again in Chapter 7).

Nevertheless, as Darwin argued eloquently in *The Origin of Species*, the general features of the fossil record provide strong evidence for evolution. The discoveries of palaeontologists since his day have reinforced his arguments. In the first place, many examples of intermediate forms have been discovered, connecting groups that were formerly thought to be separated by unbridgeable gaps. The fossil bird-reptile *Archaeopteryx*, discovered shortly after the publication of *The Origin of Species*, is perhaps the most famous of these. *Archaeopteryx* fossils are rare (only six specimens exist). They come from Jurassic limestone from about 120 million years ago that was laid down in a large lake in Germany. These creatures show a mosaic of characteristics, some resembling those of modern birds, such as wings and feathers, and others like those of reptiles, such as a toothed jaw (instead of a beak) and a long tail. Many details of their skeletons are indistinguishable from those of a contemporary group of dinosaurs, but *Archaeopteryx* differs from them, as it could clearly fly. Other fossils linking birds and dinosaurs have subsequently been found, and it has recently been shown that dinosaurs with feathers existed before *Archaeopteryx*. Other important intermediates include fossil mammals from the Eocene (about 60 million years ago), with forelimbs and reduced hindlimbs adapted to swimming. These link modern whales to animals that belong to the group of cloven-hoofed herbivores that includes cows and sheep.



Humans are an excellent example of gaps in the record being filled as more research is done. No fossil remains connecting apes and humans were known at the time of publication of Darwin's 1871 book on human evolution, *The Descent of Man*. Darwin argued on the basis of anatomical similarities that humans were most closely related to gorillas and chimpanzees, and had therefore probably originated in Africa from an ancestor that also gave rise to these apes. A whole series of fossil remains have since been found and accurately dated by the methods described earlier, and new fossils continue to be found. The nearer in time to the present, the more similar are the fossils to modern humans (Figure 10); the earliest fossils that can be assigned clearly to *Homo sapiens* date from only a few hundred thousand years ago. In agreement with Darwin's inferences, early human evolution probably took place in Africa, and it seems likely that our relatives first entered Eurasia about 1.5 million years ago.

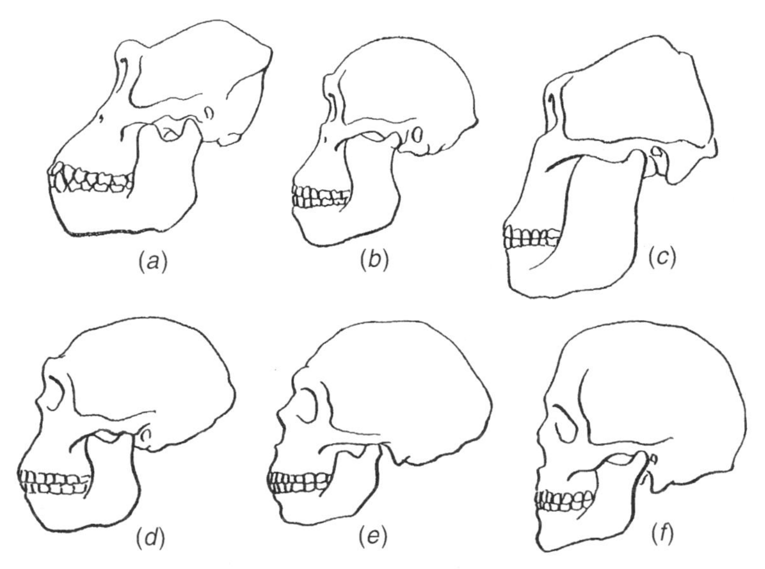


**10. Skulls of some human ancestors and relatives. A. Female gorilla. B and C. Fossils of two different species of one of the earliest human relatives, *Australopithecus*, from about 3 million years ago. D. Fossil of an intermediate between *Australopithecus* and modern humans called *Homo erectus*, from about 1.5 million years ago. E. A fossil Neanderthal human, *Homo neanderthalensis*, from about 70,000 years ago. F. Modern human, *Homo sapiens*.**



There are also cases of almost completely continuous temporal sequences of fossils, in which it seems certain that we have a record of change in a single evolving lineage. The best examples come from studies of the results of drilling down into deposits at the bottom of the sea, from which long rock columns can be recovered. This allows very fine-scaled time separation between successive samples of the microorganisms whose innumerable fossilized skeletons form the body of the rock. Careful measurements of the shapes of the skeletons of creatures such as foraminiferans, which are single-celled marine animals, allows characterization both of the averages and levels of variability of successive populations over a long period of time (Figure 11).

Even if there were no graded intermediates in the fossil record, the general features of the record are barely comprehensible except in

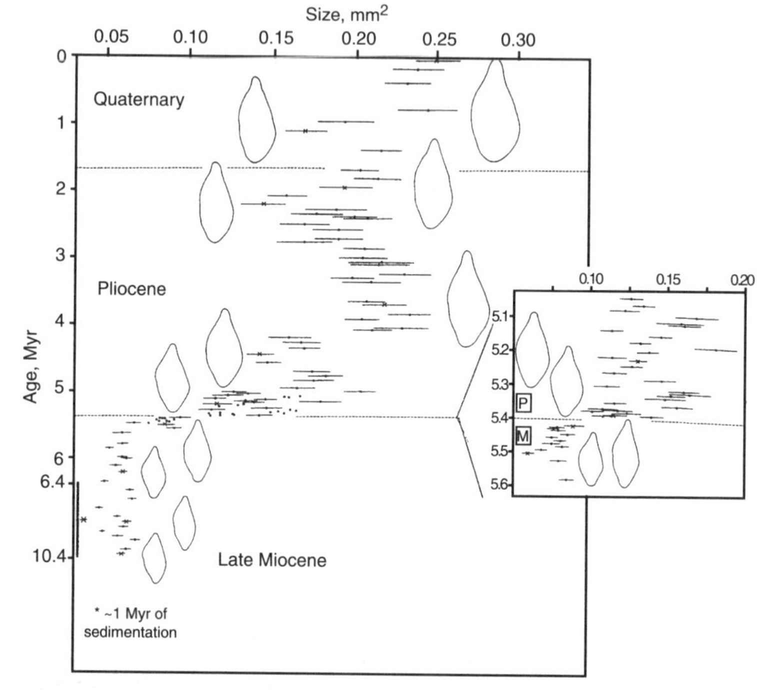


**11. Gradual evolutionary change in a series of fossils. The figure shows the means and ranges of body sizes in samples of a fossil foraminiferan (*Globorotalia tumida*), a single-celled marine shelled animal. Size changes gradually in this lineage, except for two apparent discontinuities. At the boundary between the late Miocene and Pliocene eras, a more detailed set of fossils (inset) shows that the discontinuity observed with the coarser set of fossils almost entirely reflects an episode of very rapid change, since the ranges of most successive samples overlap each other. For the discontinuity just before 4 million years ago, there are currently no fossil data.**

the light of evolution. Although the fossil record before the Cambrian era is fragmentary, there is evidence for the remains of bacteria and related single-celled organisms going back more than 3.5 billion years. Much later on, there are remains of more advanced (eukaryote) cells, but still no evidence for multicellular organisms. Organisms made up of simple clusters of cells appear only about 800 million years ago (MYA), at a time of environmental crisis when the Earth was largely covered with ice. About 700–550 MYA, there is evidence for soft-bodied, multicellular animal life.

As already mentioned, animal remains associated with hard skeletons only become abundant in the Cambrian rocks, about 550 MYA. By the end of the Cambrian, around 500 MYA, there is evidence for nearly all major animal groups, including primitive fish-like vertebrates that lack jaws, resembling modern lampreys.

All life until this time is associated with marine deposits, and the only plant remains are algae, which lack the vessels that multicellular land plants use for fluid transport. By 440 MYA, there is evidence for freshwater life, followed by fossil spores that imply the existence of the first land plants; shark-like fish with jaws appear in the sea. In the Devonian (400–360 MYA), freshwater and land remains become much more common and diverse. There is evidence for primitive insects, spiders, mites, and millipedes, as well as simple vascular plants and fungi. Jawed fish with bony skeletons become common, including lobe-finned fishes similar in structure to the first salamander-like amphibians that appear at the end of the Devonian. These are the first land vertebrates.



During the next division of the geological record, the Carboniferous (360–280 MYA), land life-forms become abundant and diverse. The coal deposits, which give this period its name, are the fossilized remains of tree-like plants that grew in tropical swamps, but these are similar to contemporary horse-tails and ferns and are unrelated to modern conifers or deciduous trees. Remains of primitive reptiles, the first vertebrates to become fully independent of water, are found at the end of the Carboniferous. In the Permian (280–250 MYA), there is a great diversification of reptiles; some of these have anatomical features that increasingly come to resemble those of mammals (the mammal-like reptiles). Some of the modern groups of insects, such as bugs and beetles, appear.

The Permian ends with the largest set of extinctions seen in the fossil record, in which some previously dominant groups such as trilobites disappear completely, and many other groups are nearly wiped out. In the recovery that follows, a variety of new forms appear, both on the land and in the sea. Plants similar to modern conifers and cycads appear in the Triassic (250–200 MYA). Dinosaurs, turtles, and primitive crocodiles appear; right at the end, the first true mammals are found. These are distinguished from their precursors by having a lower jaw consisting of a single bone connected to the skull directly (the three bones that form this connection in reptile skulls have evolved into the small internal bones of the mammalian ear, see Chapter 3, p. 15). Bony fishes similar to modern forms appear in the sea. In the Jurassic (200–140 MYA), the mammals diversify somewhat, but life on land is still dominated by the reptiles, especially dinosaurs. Flying reptiles and *Archaeopteryx* appear. Flies and termites appear for the first time, as do crabs and lobsters in the sea. Only in the Cretaceous (140–65 MYA) did flowering plants evolve – the last major group of organisms to evolve. All major modern groups of insects are found by this time. Pouched mammals (marsupials) appear in the middle of the Cretaceous, and forms similar to modern placental mammals are found towards its end. Dinosaurs are still abundant, though in decline at the end of the period.

The Cretaceous ends with the most famous of the major extinction events, associated with the impact of an asteroid that landed in the Yucatan peninsula of Mexico. All the dinosaurs (except birds) disappear, along with many other forms once common on the land and in the sea. This is followed by the Tertiary period, which extends until the beginning of the great Ice Ages, about 2 MYA. During the first divisions of the Tertiary, between 65 and 38 MYA, the main groups of placental mammals appear. At first, these are mostly similar to modern insectivores such as shrews, but some become fairly distinct by the end of this period (whales and bats are recognizable, for instance). Most of the main groups of birds are found, as well as modern types of invertebrates, and all major flowering plant groups except grasses. Bony fishes of an essentially modern type are abundant. Between 38 and 26 MYA, grasslands appear, associated with grazing horse-like animals with three toes (instead of the single toe of modern horses). Primitive apes also appear. Between 26 and 7 MYA, prairie grasslands are widespread in North America, and horses with short lateral toes and high-crowned teeth adapted for grazing are found. Various ungulates, such as pigs, deer, and camels appear, together with elephants. Apes and monkeys become more diverse, especially in Africa. Between 7 and 2 MYA, marine life has an essentially modern aspect, although many forms living then are now extinct. The first remains of creatures with some distinctively human features appear in this period. The end of the Tertiary, from 2 MYA to 10,000 years ago, sees a series of Ice Ages. Most animals and plants are essentially modern in form. Between the end of the last Ice Age 10,000 years ago, and the present, humans become the dominant land animal, and many large mammal species become extinct. There is some fossil evidence for evolutionary change over this period, such as the evolution of dwarf forms of various large mammal species on islands.

The fossil record thus suggests that life originated in the sea over 3 billion years ago, and that for more than a billion years only single-celled organisms related to bacteria existed. This is exactly what is expected on an evolutionary model; the evolution of the machinery needed to translate the genetic code into protein sequences, and the complex organization of even the simplest cell, must have required many steps, the details of which almost defy our imagination. The late appearance in the record of clear evidence for eukaryote cells, with their substantially more complex organization compared with prokaryotes, is also consistent with evolution. The same applies to multicellular organisms, whose development from a single cell requires elaborate signalling mechanisms to control growth and differentiation: these could not have evolved before single-celled forms existed. Once simple multicellular forms evolved, it is understandable that they rapidly diversified into numerous forms, adapted to different modes of life, as occurred in the Cambrian. We shall discuss adaptation and diversification in the next chapter.

The fact that life was exclusively marine for an immense period is also understandable from an evolutionary perspective. Early in the Earth's history, the geological evidence shows that there was very little oxygen in the atmosphere. The consequent lack of protection from ultra-violet radiation by atmospheric ozone, which is formed from oxygen, would have prohibited life on land or even in fresh water. Once sufficient oxygen had built up as a result of the photosynthetic activities of early bacteria and algae, this barrier was removed, and the possibility of the invasion of the land opened up. There is evidence for an increase in atmospheric oxygen levels during the period leading up to the Cambrian, which may have permitted the evolution of larger and more complex animals. Similarly, the appearance of fossils of flying insects and vertebrates after the emergence of life on land makes sense, since it is unlikely that true flying animals could evolve from purely aquatic forms.



The recurrent phenomenon of the emergence of abundant and diverse forms of life, followed by their wholesale extinction (as with the trilobites and dinosaurs) or their reduction to just one or a few surviving forms (like the coelacanths) also makes sense in terms of evolution, whose mechanisms have no foresight and cannot guarantee that their products can survive sudden large environmental changes. Similarly, the rapid diversification of groups after the colonization of a new habitat (as in the invasion of the land), or after the extinction of a dominant rival group (as with the mammals after the disappearance of the dinosaurs), is expected on evolutionary principles.

The interpretation of the fossil record in terms of biological knowledge therefore follows the same principle of uniformitarianism that is applied by geologists to the history of the structure of the Earth. The fossil evidence might have shown patterns that falsify evolution. The great evolutionist and geneticist J. B. S. Haldane is alleged to have answered the question of what observation would cause him to abandon his belief in evolution by saying: 'A pre-Cambrian rabbit'. So far, no such fossil has been found.

## Patterns in space

Another important body of facts that makes sense only in terms of evolution comes from the distribution of living creatures over space rather than time, as described by Darwin in two of the fifteen chapters of the *The Origin of Species*. One of the most striking examples of this involves the flora and fauna of oceanic islands, such as the Galapagos and Hawaiian islands, which geological evidence shows were formed by volcanic action and were never connected to a continent. According to the theory of evolution, the present-day inhabitants of such islands must be the descendants of individuals who were able to cross the vast distances separating the newly formed islands from the nearest inhabited land. This puts several restrictions on what we are likely to see. First, the difficulty of colonization of a remote piece of newly formed land means that few species will be able to establish themselves. Second, only types of organism that have characteristics that enable them to cross hundreds or thousands of miles of ocean can become established. Third, even in the groups that are represented, there will be a highly random element to which species are present, because of the small number of species that arrive on the islands. Finally, evolution on such remote islands will produce many forms that are found nowhere else.

These expectations are strikingly verified by the evidence. Oceanic islands do indeed tend to have relatively few species in any major group, compared with continents or offshore islands with comparable climates. The types of organisms found on oceanic islands, before human introductions, are wildly unrepresentative of those found elsewhere. For example, reptiles and birds are usually present, whereas terrestrial mammals and amphibians are consistently missing. In New Zealand, there were no terrestrial mammals before human occupation, though there were two species of bats. This reflects the ability of bats to cross large bodies of salt water. The rampant spread of many species after human introduction shows clearly that the local conditions were not unsuitable for their establishment. But even among the major types of animals and plants that are present, whole groups are often missing, whereas others are disproportionately represented. Thus, on the Galapagos islands, there are just over 20 species of land birds, of which 14 are finches, the famous finches described by Darwin in his account of his voyage round the world in *HMS Beagle*. This is quite unlike the situation elsewhere, in which finches form only a small fraction of the land bird fauna. It is exactly what one would expect if there were only a small number of species of original bird colonists, one of which was a species of finch that became the ancestor of the present-day species.



As this view would predict, oceanic islands provide many examples of forms that are unique to them, but nevertheless show affinities to mainland species. For example, 34% of the plant species found on the Galapagos islands are present nowhere else. Darwin's finches have a much greater variety of beak sizes and shapes than is usual among finches (which are normally seed-eaters with large, deep beaks), and these are clearly adapted to different modes of food gathering (Figure 12). Some of these are highly unusual, such as the habit of the sharp-beaked ground finch *Geospiza difficilis* of pecking the rear ends of nesting seabirds and drinking their blood. The woodpecker finch *Cactospiza pallida* uses twigs or cactus spines to extract insects from dead wood. Even more spectacular examples of rampant evolution are found on other groups of oceanic islands. For instance, the number of species of the fruitfly *Drosophila* on Hawaii exceeds that found in the rest of the world, and they are amazingly diverse in body size, wing patterns, and feeding habits.

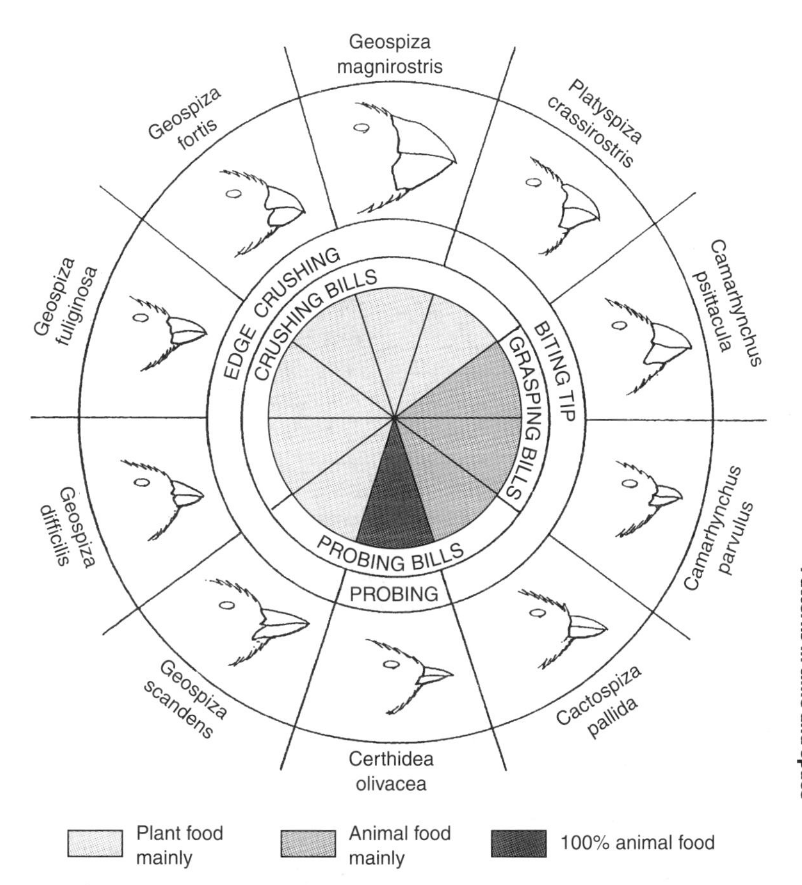


**12. The beaks of Darwin's finches, showing the differences in size and shape between species with different diets.**

These observations are explicable if the colonist ancestors of these island species found themselves in environments free from established competitor species. This situation would permit the evolution of traits that adapted the colonists to new ways of life, and allowed diversification of an ancestral species into several descendant species. Despite the unusual modifications of structure and behaviour found in Darwin's finches, studies of their DNA, by the methods described in Chapters 3 and 6, show that these species have a relatively recent origin about 2.3 million years ago, and are closely related to species on the mainland (Figure 13).

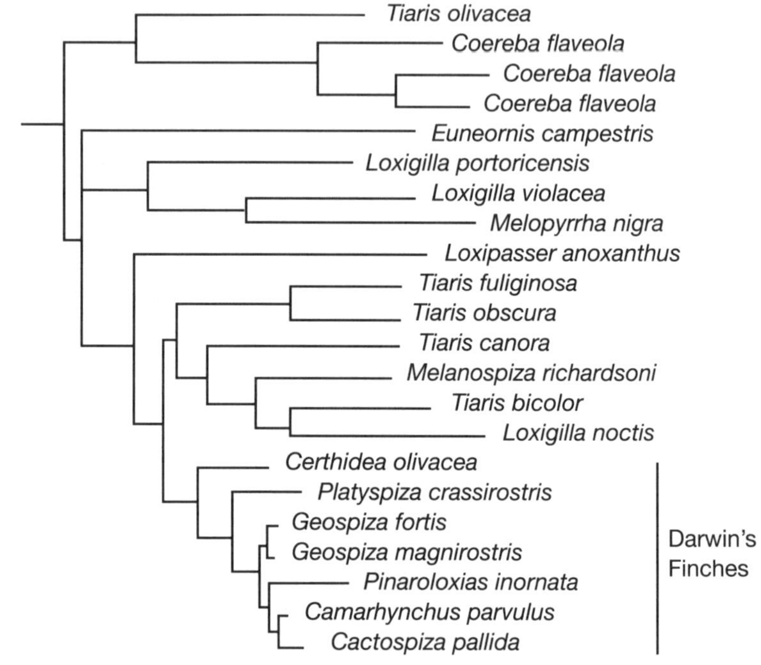

**13. Phylogenetic tree of Darwin's finches and their relatives. The tree is based on differences among different species in the DNA sequences of a gene in their mitochondria. The lengths of the horizontal branches in the tree indicate the amounts of differences between species (ranging from 0.2% between the closest species to 16.5% between the most different). The tree shows that the Galapagos species form a cluster clearly having a common ancestor, and that they all have similar sequences of this gene, consistent with this ancestor being quite recent. In contrast, the other related species of finch differ much more from one another.**

As Darwin wrote, when describing the inhabitants of the Galapagos islands in *The Origin of Species*:

> Here almost every product of the land and of the water bears the unmistakable stamp of the American continent. There are twenty-six landbirds; and twenty-five of these are ranked by Mr. Gould as distinct species, supposed to have been here created here; yet the close affinity of most of these birds to American species in every character, in their habits, gestures and tones of voice, was manifest.

So it is with the other animals, and with nearly all the plants, as shown by Dr. Hooker in his admirable memoir on the Flora of this archipelago. The naturalist, looking at the inhabitants of these volcanic islands in the Pacific, distant several hundred miles from the continent, yet feels that he is standing on American land. Why should this be so? why should the species which are supposed to have been created in the Galapagos Archipelago, and nowhere else, bear so plainly a stamp of affinity to those created in America? There is nothing in the conditions of life, in the geological nature of the islands, in their height or climate, or in the proportions in which the different classes are associated together, which resembles closely the conditions of the South American coast; in fact there is a considerable dissimilarity in all these respects.

The theory of evolution, of course, provides the answer to these questions; research on island life over the past 150 years has amply confirmed Darwin's remarkable insights.

# Chapter 5
# Adaptation and natural selection

## The problem of adaptation

An important task of the theory of evolution is to account for the diversity of living organisms within the hierarchical organization of resemblances between them. In Chapter 3, we emphasized resemblances between different groups, and how they make sense in terms of Darwin's theory of descent with modification. The second essential part of evolutionary theory is to provide a scientific explanation for the 'adaptation' of living organisms: their appearance of good engineering design, and their diversity in relation to their different ways of living. This will require the longest chapter in this book.

There are innumerable remarkable examples of adaptations, and we will just mention a few to illustrate the nature of the problem. The diversity of different kinds of eyes alone is astonishing, and yet makes sense in relation to the environments in which different animals live. Eyes for seeing underwater are different from those for seeing in air, and the eyes of predators have special adaptations to break the camouflage of prey that have evolved to be difficult to see. Many underwater predators that eat transparent marine animals have eyes with special contrast-increasing systems, including ultra-violet vision and polarized light vision. Other well-known adaptations are the hollow bones of birds' wings, with internal

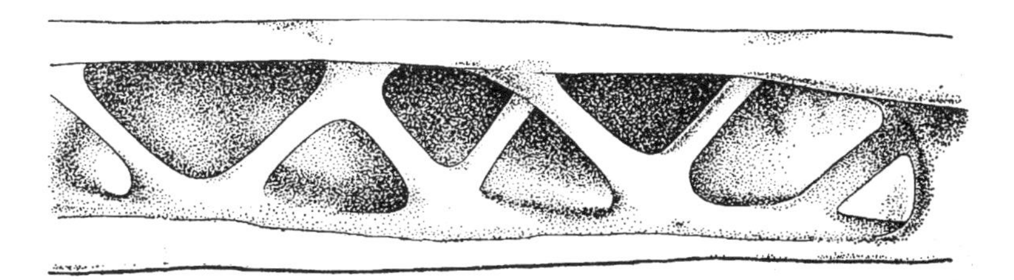

**14. A hollow bone of a vulture's wings, with its internal strengthening struts.**

struts resembling those in aircraft wings (Figure 14), or the wonderful construction of animal joints, whose surfaces allow the moving parts to move smoothly over one another.

Many other examples are provided by animal adaptations related to their different ways of feeding, and by reciprocal adaptations of the organisms that they feed on. Butterflies have long tongues to reach deep down into flowers and suck up nectar; reciprocally, flowers have high visibility to insects, and advertise themselves by scents, as well as rewarding visitors by nectar. Frogs and chamaeleons have long tongues that can shoot out and capture insect prey on their sticky tips. Many animals have adaptations to help them escape from predators, and the appearance of such animals depends on where they live. The silvery colour of many fish species makes them difficult to see in the water, but few land animals have this coloration. Some animals have cryptic coloration, with extraordinarily close mimicry of leaves or twigs, or of other poisonous or stinging species.

Adaptations are recognizable in many details of animals', plants', and microbes' lives, at every level, down to the cellular machinery and its controls (which we described in Chapter 3). For instance, cell division and cell movements are powered by tiny motors made of protein molecules. Proof-reading of newly produced DNA occurs when the genetic material is copied while making a new cell, reducing the frequency of harmful mutations several-thousand-fold. Protein complexes in cell surfaces selectively allow entry of some chemicals, but prevent others getting in. In nerve cells, these are used to control the flow of electrically charged metal atoms across the cell surface, generating the electrical signals used in transmitting information along the nerves. The behaviour patterns of animals are the ultimate outcome of the patterns of activity of their nerves, and are clearly adapted to their ways of life. For instance, in birds, nest parasites like cuckoos remove the host species' eggs or young from the nest, leaving the hosts to raise their young. In turn, the host species adapt by becoming more vigilant. Ants that grow fungus 'gardens' have evolved behaviours including weeding out spores of fungi contaminating their decaying leaves. Even the rate of ageing is adapted to the environment an animal or plant lives in, as we shall explain in Chapter 7.

Before Darwin and Wallace, such adaptations appeared to require a Creator. There seemed no other way to account for the astonishing detail and apparent perfection of many aspects of living organisms, just as the complexity of a watch could not be a purely natural production. The absence of any other explanation was the main support for the *Argument from Design* developed by 18th-century theologians to 'prove' the existence of a Creator, and the term *adaptation* was introduced to describe the observation that living things have structures that seem to be useful to them. It is important to understand that describing these as adaptations poses a question. It was a valuable contribution to our understanding of life to see that adaptations demand an explanation.

There is no doubt that animals and plants differ from other naturally produced things, such as rocks and minerals, as we acknowledge in the game 'animal, vegetable, or mineral'. But the Argument from Design overlooks the possibility that there could be natural processes, in addition to those that produce minerals and rocks, mountains and rivers, which can account for living creatures as complex natural productions, without the need for a Designer. The biological explanation of the origin of adaptation replaces the idea of a Designer, and is central to post-Darwinian evolutionary biology. In this chapter, we describe the modern theory of adaptation and its biological causes and basis. This is based on the theory of natural selection, which we outlined in Chapter 2.

## Artificial selection and heritable variation

A first, very pertinent observation, strongly emphasized by Darwin, is that the modification of organisms by humans is regularly possible, and can produce the same appearance of design that we see in nature. This is routinely achieved by *artificial selection*, or selective breeding from animals and plants with desirable characters. Very striking changes can be produced over a time-frame that is short on the scale of the fossil record of evolution. For example, we have developed many different strains of cabbages, including strange ones like the cauliflower and broccoli, which are mutants that cause monstrous flowers forming a large head, and ones like the brussels sprout in which leaf development is abnormal (Figure 15A). Similarly, many breeds of dogs have been bred by humans (Figure 15B), with differences very like those observed between different species in nature, as Darwin pointed out. However, although all *Canis* species (including coyotes and jackals) are close relatives and can interbreed, dog breeds are not domestications of different wild dog species, but have been produced over the past few thousand years (several hundred dog generations) by artificial selection from a single common ancestral species, the wolf. The DNA sequences of dog genes are essentially a subset of wolf sequences, but coyotes (whose ancestor is believed from fossils to have separated from wolves' ancestors a million years ago) are about three times as different from either dogs or wolves as the most different dog/wolf comparison. The differences among dogs in their sequences of the same gene, differences which presumably developed after dogs separated from wolves, can be used to tell how long ago that separation happened (see Chapter 3). The conclusion is that dogs separated from wolves much longer ago than 14,000 years, the date suggested by archaeological records, but not more than 135,000 years ago.

A

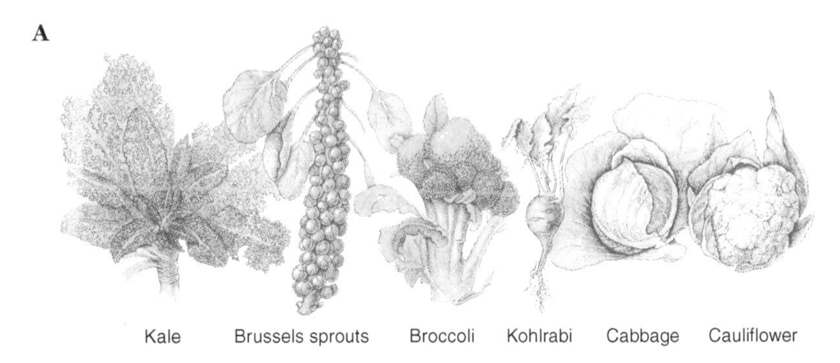


B

**15. A. Some of the diversity of cultivated varieties of cabbage. B. Differences in the sizes and shapes of two breeds of dog.**

The success of artificial selection is possible because heritable variation exists within populations and species (the slight differences between normal individuals which we described in Chapter 3). Even without any understanding of inheritance, people have bred from animals and plants that had characteristics they liked or found useful, and over enough generations this process has generated strains of animal and plant species that differ greatly from one another, and from the ancestral forms that were originally domesticated. This shows clearly that individuals within domesticated species must have been different from one another, and that many differences can be passed from parents to their offspring, that is they are heritable. If differences were merely due to the way the animals or plants were treated, selective breeding and artificial selection would have no effect on the next generation. Unless some of the differences were heritable, the breed could improve only by improved husbandry.

Every imaginable kind of character can vary heritably. The different breeds of dog differ, as is well known, not just in appearance and size, but also in mental traits such as character and disposition, some tending to be friendly, while others are fierce and suitable as guard dogs. They differ in their interest in scents, and in their inclination to fetch and carry or to swim, and in intelligence. They differ in the diseases to which they are susceptible, as in the well-known case of Dalmatians being prone to gout. They even differ in the ageing process, with some breeds, such as the Chihuahua, having surprising longevity (their life-span is almost as long as that of cats), while others, such as the Great Dane, live only about half as long. Although all these characteristics are, of course, affected by environmental circumstances such as good care and treatment, they are strongly influenced by heredity.

Similar heritable differences are known in many other domesticated species. To take another example, the qualities of different apple varieties are heritable differences. They include adaptations to different human needs such as early or late harvesting, suitability for cooking or eating, and to the differing climates of different countries. Just as in the case of dogs, other evolutionary processes have gone on in apples at the same time as human selection, and perfection is never attained for all desirable traits. For instance, Coxes are a particularly flavoursome apple, but are highly susceptible to disease.

## Kinds of heritable variation

The success of artificial selection is very strong evidence that many kinds of character differences in animals and plants are heritable. There are also many genetic studies showing heritable variation for the characteristics of a wide range of organisms in nature, including many different species of animals, plants, fungi, bacteria, and viruses. Variation originates by well-understood processes of random mutation in the DNA sequences of genes, similar to those that produce human genetic disorders (Chapter 3). Most mutations are probably deleterious, like the genetic disorders of humans and farm animals, but advantageous mutations do sometimes occur. Such mutations have led to the resistance of animals to disease (such as the evolution of myxomatosis resistance in rabbits). They are also responsible for a major problem today, pests evolving resistance to chemicals (including resistance of rats to warfarin, or of worms parasitic in humans and farm animals to antihelminthic chemicals, insecticide resistance in mosquitoes, and antibiotic resistance in bacteria). Because of their importance to human or animal welfare, many cases are understood in great detail.



Heritable differences are also well known in humans. Variation may take the form of 'discrete' character differences, such as eye and hair colour, as already mentioned. These are variants controlled by differences in single genes, and unaffected by environmental circumstances (or altered only slightly, for instance when a fair-haired person's hair is bleached by the sun). Common variants like these are called *polymorphisms*. Conditions such as colour blindness are also simple genetic differences, but are much rarer variants in human populations. Even behavioural characters may be heritable. Whether fire ant colonies have single or multiple queens seems to be controlled by a difference in a single gene for a protein that binds a chemical involved in recognition of other individuals.

'Continuous' variation is also very evident for many characters in populations, for example the gradations of height and weight among people. This kind of variation is often markedly affected by environmental conditions. The increasing height of successive generations during the 20th century, seen in many different countries, is not due to genetic changes but to changed conditions of life, including better nutrition and fewer serious illnesses during childhood. Nevertheless, there is also some degree of genetic determination for such characters in human populations. This is known from studies of identical and non-identical twins. Non-identical twins are ordinary siblings that happen to be conceived at the same time, and they differ as much as any siblings, but identical twins come from a single fertilized egg that splits into two embryos, and are genetically identical. Greater resemblances between identical than non-identical twins have been documented for many characteristics, which must be due to their genetic similarity (care must, of course, be taken that the identical twins are not treated more alike than non-identical pairs – for instance, only same-sex pairs of both kinds of twins should be studied). Despite the important environmental influences that clearly often exist, this and other kinds of evidence consistently reveal some degree of heritable basis for variation in many characteristics, including aspects of intelligence. Heritable variation has been documented in many organisms, and for all kinds of characteristics. Even an animal's place in the dominance hierarchy, or pecking order, can be heritable; this has been demonstrated in chickens and in cockroaches. The amount of continuous genetic variability can be measured from resemblances between relatives of different degrees. This is useful in animal and crop plant breeding, and allows breeders to predict the characteristics, such as milk yield of cows, that offspring of different parents will have, and thus to plan their breeding.

Genetic differences boil down to differences in the 'letters' in the DNA. These often leave the amino acid sequences of proteins unchanged. When the DNA sequences of the same gene are compared between different individuals, differences are seen, though usually fewer than when sequences are compared between different species (such comparisons were discussed in Chapter 3, see Figure 8). For example, copies of the gene for glucose-6-phosphate dehydrogenase, mentioned in Chapter 3, one from each of a set of individuals, might be compared. There may be no differences (so no diversity). If some individuals in the population have a variant sequence of the gene, the difference will show up in some of the comparisons. This is called molecular polymorphism. Geneticists measure such diversity by the fraction of the letters in the DNA sequence that vary between individuals in the population. In the human species, it is usually found that fewer than 0.1% of the DNA letters differ when we compare the same gene's sequence between different people (compared with generally around 1% of the letters being different when a gene's sequences are compared between a human and a chimpanzee). Variation is higher in some genes and lower in others, and, as one might expect, variation is generally higher in the presumably less important regions of the genome that do not code for proteins than in the coding parts of genes. Humans are rather lacking in variability compared with most other species. For example, DNA polymorphism is much more common in maize (more than 2% of its DNA letters are variable).

The distribution of variability within a species can give us useful information. When dogs are bred for different characteristics, breeds are developed that are rather uniform in their characteristics. This is due to strict pedigree rules, which control matings and forbid 'gene flow' between breeds. A characteristic that is desired in one breed, such as fetching, is thus well developed in that breed only, and separate breeds tend to diverge from each other. This isolation between breeds is unnatural, and dogs of different breeds will happily mate and produce healthy young. Much of the variability of dogs is accordingly between breeds. Many natural species live in different, geographically separated populations, and, as one might expect, the amount of diversity in such species as a whole is greater than within a single population, because there are differences between populations. For example, certain blood groups are more common in some human races than in others (see Chapter 6), and the same is true for many other genetic variants. However, in humans and many other species in nature, the differences between populations are very slight in comparison with the diversity within populations, unlike the situation for dog breeds. The difference is because humans move freely between populations. An important implication of these genetic results is that human races are distinguished by a small minority of the genes in our genomes, most of our genetic make-up having a similar range and heterogeneity of variants worldwide. Increased mobility in the modern world is quickly reducing any differences between populations.



## Natural selection and fitness

A fundamental idea in the theory of evolution under natural conditions is that some heritable character differences affect survival and reproduction. For instance, just as race horses have been selected for speed (by breeding from winners and their relatives), so antelopes have been naturally selected for speed, because the individuals that breed and contribute to the future of their species are those that did not get eaten by predators. Darwin and Wallace realized that this kind of process could explain adaptation to natural conditions. Our ability to modify animals and plants by artificial selection depends on this characteristic having a heritable basis. Provided that there are heritable differences, successful individuals in the wild will likewise pass their genes (and thus often their good characteristics) to their offspring, which will, in turn, possess the adaptive characters, such as speed.

For brevity, and to allow one to think in general terms, the word *fitness* is often used in biological writing to stand for overall ability to survive and reproduce, without the need to specify which characters are involved (just as we use the term 'intelligence' to mean a variety of different abilities). Many different aspects of organisms contribute to fitness. For instance, speed is just one feature affecting antelope fitness. Alertness and the ability to detect predators are also important. Mere survival is not enough, however, and reproductive abilities, such as provisioning and care of the young, are also important for fitness in animals, and the ability to attract pollinators is critical for fitness in flowering plants. The word fitness can accordingly be used to describe selection acting on a wide range of different traits. As with 'intelligence', the generality of the term 'fitness' has led to misunderstandings and disputes.

To know what characters are likely to be important for the fitness of an organism, one must understand a great deal about its biology and the environment in which it lives. The same character may give high fitness in one species, but not in another. For instance, speed is not important for fitness in a lizard that evades predators by cryptic coloration. If such a lizard lives in trees and perches on twigs, it is more important for it to be good at holding on than to run fast, and so short legs, not long ones, will be associated with high fitness. Speed is adaptive for antelopes, but staying very still, so as not to be detected by predators, is an alternative means by which many animals avoid being eaten. Other animals avoid predators by frightening them away; for example, some butterflies have eye spots in their wing patterns that can be suddenly displayed in order to alarm birds. Plants obviously cannot move, and avoid being eaten by different means, including tasting bad or being prickly. All these different characteristics may increase the survival and/or reproduction of the organisms, and hence their fitness.

Given genetic variability for many characters, and environmental differences, natural selection will inevitably operate, and the genetic make-up of populations and species will change over time, as we showed in Chapter 2. Changes are often slow in terms of years, because it takes many generations for a genetic variant that is rare to become the majority type in the population. In animal and plant breeding, severe selection often occurs (for example, when diseases wipe out most of a herd or a plant crop), but changes still take many years. It is estimated that maize was domesticated about 10,000 years ago, but modern giant corn cobs are a quite recent development. Despite the slowness of evolutionary change in terms of years, natural selection can produce rapid changes on the timescale of the fossil record. Advantageous traits can spread throughout a population from a very low starting frequency in less time than that between successive layers in the geological strata (usually at least several thousand years, see Chapter 4).

Even though we often may not see it happening, because of its slowness in terms of the time-scale of our lives, natural selection never stops. Even humans are still evolving. For instance, our diet differs from that of our ancestors, and our teeth can function quite well on soft modern foods even if they are not very strong. The high sugar content of many modern foods leads to tooth decay, and potentially to abscesses that can be fatal, but there is no longer very pronounced natural selection for strong teeth, because dental care can solve these problems, or provide false teeth. Just as for other functions that are no longer used intensively, changes are to be expected, and our teeth could one day become vestigial. They are already smaller than those of our close relatives, the chimpanzees, and there is no reason why they should not become smaller still. Excess sugar in the diet has also led to an increasing frequency of late-onset diabetes in human populations, with high mortality for sufferers. In the past, this disease was largely confined to people past childbearing age, but the age of onset is becoming steadily earlier. There is therefore a new, probably intense, selection pressure to change our metabolism so as to tolerate our changed diet. In Chapter 7, we will show how changes in human life are leading to the evolution of greater longevity.

The concept of fitness is often misunderstood. When biologists try to illustrate the meanings of this term, they often use examples that correspond with our everyday use of the word fitness, such as the speed of antelopes. There is less danger of confusion if we think of characteristics like the lightweight bones of birds, with their hollow centres and strengthening cross-struts (Figure 14). The theory of natural selection accounts for such apparently well-designed structures by pointing out that, when flight was evolving, lighter-boned individuals would have had slightly higher chances of survival than others. If their descendants inherited lighter bones, the characteristic would increase in its representation in the population over the generations. This is just the same as artificial selection by breeders of the fastest dogs, which has given all greyhounds long, thin legs. These are mechanically more efficient than short ones, and greyhounds' legs closely resemble those of antelopes and other fast-running animals, which have evolved by natural selection. We can describe natural and artificial selection perfectly well without using the word fitness. Natural selection implies nothing more than that certain heritable variants may be preferentially passed on to future generations. Individuals carrying genes that lower their survival or reproductive success will generally not pass on those genes to the same extent as other individuals whose genes give higher survival or reproductive ability. The term fitness is merely a useful short-cut to help express briefly the idea that characteristics sometimes affect organisms' chances of surviving and/or reproducing, without having to specify a particular characteristic. It is also useful in making mathematical models of the way selection affects the genetic make-up of a population. Conclusions from these models provide a rigorous underpinning for many of the statements that we make in this chapter, but we will not describe them here.

To illustrate selection of an advantageous mutation, consider the arms race between humans and rats, in which we try to develop rat poisons, and rats evolve resistance. The rat poison warfarin kills rats because it prevents blood clotting. It binds to an enzyme needed in the metabolism of vitamin K, which is important for blood clotting and many other functions. Resistant rats were once rare, because their vitamin K metabolism is changed, reducing growth and survival. In other words, there is a *cost* of resistance. In farms and towns where warfarin is used, however, only resistant animals can survive, so there is strong natural selection, despite the cost. The resistant version of the gene has therefore spread to high frequencies in the rat population, though the cost keeps it from spreading to all members of the species. However, a recent development is the evolution of a new kind of resistance which seems to be free from the cost, and which may even be advantageous (in the absence of poison). There is thus continued evolution in response to a change in the rats' environment.

Variability and selection are very general properties of many systems, not just individual organisms. Certain components of the genetic material are maintained, not because they increase the fitness of the organisms that carry them, but because they can multiply within the genetic material itself, just like parasites in the body of their host. 50% of human DNA is thought to belong in this category. Another important situation in which natural selection drives evolutionary change within an organism occurs in cancers. Cancer is a disease in which a cell and its descendants evolve selfish behaviour and multiply, regardless of the good of the rest of the body. The disease is often caused by a mutation that increases the mutation rates of other genes (for instance, by a failure in the proof-reading system described in Chapter 3, which checks DNA sequences and prevents mutations). If mutations occur at a high frequency, some may affect cell multiplication rates, and a fast-multiplying lineage may appear. As time goes on, more and more of the cells will descend from cells carrying mutations in other genes which confer faster and faster growth, and so the cancer often becomes more aggressive. Cancer cells can also become resistant to drugs that suppress their growth. Like the well-known situation of drug-resistant HIV viruses evolving in an AIDS patient, cancer cells which acquire mutations that allow them to escape drug suppression outgrow the initial type of cells, and cause loss of remission of cancers. This is why it is often hopeless to restart drug treatment after a remission stops.

At the other extreme, there may be different rates of extinction of species with different sets of characteristics, that is there can be selection at the level of species. For example, species with large body sizes, which tend to have low population sizes and low rates of reproduction, are more vulnerable to extinction than species with small bodies (see Chapter 4). In contrast, selection between individuals of the same species often favours larger body size, probably because larger individuals have greater success in competition for food or mates. The range of body sizes that we see in a group of related species may reflect the net outcome of both types of selection. Selection on individuals within species is likely, however, to be the most important factor, since it produces the different range of body sizes in the first place, and it usually operates much faster than selection at the species level.



Selection is also important in non-biological contexts. In designing machines and computer programs, it has been found that a very efficient way to find the optimal design is to successively make small, random changes to the design, keeping versions that do the job well, and discarding others. This is increasingly being used to solve difficult design problems for complex systems. In this process, the engineer does not have a design in mind, but only the desired function.

## Adaptations and evolutionary history

The theory of evolution by natural selection explains features of organisms as a result of the successive accumulation of changes, each giving higher survival or reproductive success. What changes are possible depends on the pre-existing state of the organism: mutations can only modify the development of an animal or plant within certain limits, which are constrained by the underlying existing developmental programmes that lead to the adult organism. The results of artificial selection as practised by animal and plant breeders show that it is relatively easy to change the sizes and shapes of body parts, or to produce striking changes in superficial characters such as external coloration, as in different breeds of dog. Radical changes can easily be produced by mutations, and laboratory geneticists have no difficulty in creating strains of mice or fruitflies that differ much more from normal forms than wild species differ from each other. It is possible, for example, to produce flies with four wings instead of the normal two. These major changes, however, often severely disrupt normal development, reducing survival and fertility, and are therefore unlikely to be favoured by natural selection. They even tend to be avoided by animal and plant breeders (although such mutations have been used in developing unusual pigeon and dog breeds, where the animals' health is of lesser importance than for farmers).

For this reason, we expect that evolution will usually proceed by fairly small adjustments to what has gone before, rather than by sudden jumps to radically new states. This is particularly obvious for complex traits that depend on the mutual adjustment of many different components, such as the eye (which we discuss in more detail in Chapter 7); if one component is changed drastically, it may not function well in combination with other parts that remain unchanged. When new adaptations evolve, they will usually be modified versions of pre-existing structures, and will at first often not be the optimum functional engineering design solutions. Natural selection resembles an engineer improving machinery by tinkering with it and modifying it, rather than sitting down and planning entirely new designs. Modern screwdrivers can be suitable for precision work, with a diversity of heads suited for different purposes, but the evolutionary ancestors of screws were coarse-threaded spigots turned by a spike through a hole in one end.

While we are often astonished by the precision and efficiency of adaptations of living organisms, there are many examples of

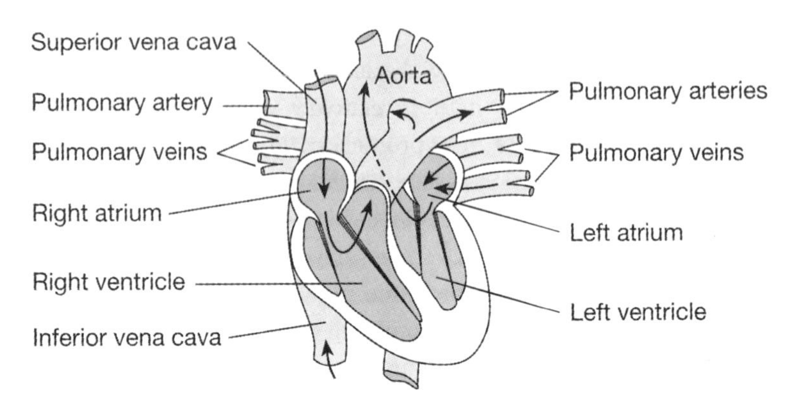


**16. The highly complex structure of the mammalian heart and its blood vessels. Note how the pulmonary artery (which delivers blood to the lungs) curves awkwardly back behind the aorta (which delivers blood to the rest of the body) and the superior vena cava (which brings blood back to the heart from the head).**



tinkering, betrayed by features that make sense only in terms of their historical origins. Painters represent angels with wings on their shoulders, allowing them the continued use of their arms. But the wings of all real flying or gliding species of vertebrates are modified forelimbs, so that pterodactyls, birds, and bats have all lost the use of their forelimbs for most of their original functions. Similarly, the design of the mammal heart and circulation has bizarre features that reflect a history of gradual modification from a system that originally functioned to pump blood from the heart round the gills of a fish, and then to the rest of the body (Figure 16). The embryonic development of the circulatory system clearly betrays its evolutionary antecedents.

Sometimes, similar solutions to a functional problem have been evolved independently in different groups, resulting in very similar adaptations that nevertheless differ considerably in detail, because of their different histories, as in the case of the wings of birds and bats. Thus, while the similarity of different organisms is often due to their being related (like ourselves and the apes), distantly related organisms living in similar circumstances can sometimes look more similar than closer ones. When morphological similarities and differences are misleading, the true evolutionary relationships can be discovered using evidence from DNA sequence similarities and differences, as explained in Chapter 3. For example, several species of river dolphins have evolved in great rivers in different parts of the world. They share some features differentiating them from open-sea species, particularly reduced eyes, because they live in turbid water, and rely more on echo-location than vision for navigation. DNA sequence comparisons show that a given species of river dolphin is more closely related to the sea-living species in its region than to river dolphins elsewhere. It makes sense that similar environments lead to similar adaptations.

Despite the similarities, natural selection differs from human design processes in several ways. One difference is that evolution has no foresight; organisms evolve in response to prevailing environmental conditions at one time, and this may result in features which lead to their extinction when conditions change radically. As we show later in this chapter, sexual competition among males can lead to structures that severely reduce their survival ability; it is quite possible that in some cases an unfavourable environmental change could further reduce survival to such a point that the species could not maintain itself, as has been suggested for the extinct Irish Elk, with its enormous antlers. Long-lived organisms often evolve very low fertility, as in the case of large birds of prey such as condors that only produce one offspring every other year (we discuss this further in Chapter 7). Such populations can do well if the environment is favourable, and there is low annual mortality of breeding adults. However, if the environment deteriorates and mortality increases, for example because of human disturbances, this may cause a rapid decline in population number. This is happening at the present time to many species, and has caused the extinction even of species that were once very abundant. For instance, the slow-breeding passenger pigeon of the USA was hunted to extinction in the 19th century despite originally having populations of tens of millions. Species which evolve to occupy an extremely specialized type of habitat are also vulnerable to extinction if that habitat disappears due to environmental change; for example, pandas in China are under threat because they breed slowly and depend on a type of bamboo found only in certain mountainous regions, which are now being logged.

Natural selection also does not necessarily produce perfect adaptation. In the first place, there may not be time to adjust every aspect of a piece of biological machinery to its best-functioning state. This is particularly likely to be true when selection pressures result from interactions between a pair of species, such as a host and a parasite. For example, an improvement in the ability of the host to resist infection increases the pressure of selection on the parasite to overcome this resistance, forcing the host to evolve new resistance measures, and so on, so that is there is an 'evolutionary arms race'. In such situations, neither partner can remain perfectly adapted for very long. Despite the wonderful ability of our immune system to combat bacterial and viral infections, we remain vulnerable to newly evolved strains of influenza and cold viruses. Second, the tinkering aspect of selection, modifying what has gone before, constrains what selection can achieve, as we have just mentioned. It seems absurd from a design point of view that the nerves that carry information from the light-sensitive cells of the vertebrate eye are in front of, rather than behind, the light-sensitive retinal cells, but this is a consequence of the way this part of the eye develops as an outgrowth of the central nervous system (the octopus eye resembles that of mammals, but has a better arrangement, with the light-sensitive cells in front of the nerves). Third, an improvement in one aspect of the functioning of a system may have a cost with respect to some other function, as mentioned in relation to warfarin resistance. This can prevent improved adaptation. We will mention other examples later in this chapter, and in Chapter 7, when we discuss ageing.

## Detecting natural selection

Darwin and Wallace argued that natural selection is the cause of adaptive evolution without knowing examples of selection operating in nature. Over the last 50 years, many cases of natural selection have been detected in action and studied in detail, immeasurably strengthening the evidence for its key role in evolution. We have space for only a few examples. A very important kind of natural selection acting today is causing ever-increasing antibiotic resistance in bacteria. This is an example of evolutionary change that is intensively studied, because it endangers our lives, and occurs fast and (unfortunately) very repeatedly. On the day we were writing this, the headlines in the newspaper were about methicillin-resistant *Staphylococcus* in Edinburgh's Royal Infirmary. Whenever an antibiotic is widely used, resistant bacteria are soon found. Antibiotics were first widely used in the 1940s, and concerns about resistance were soon being raised by microbiologists. In 1955, an article in the *American Journal of Medicine*, aimed at doctors, wrote that the indiscriminate use of antibiotics: 'is fraught with the risk of selecting resistant strains', and in 1966 (when people had not changed their behaviour), another microbiologist wrote: 'is there no way to generate sufficient general concern so that antibiotic resistance can be attacked?'

The speedy evolution of antibiotic resistance is not surprising, because bacteria multiply fast and are present in enormous numbers, so that any mutation that can make a cell resistant is sure to occur in a few bacteria in a population; if the bacteria are able to survive the change to their cell functions caused by the mutation and to multiply, a resistant population can rapidly build up. One might hope that resistance will be costly for bacteria, as was initially true for warfarin resistance in rats, but as in rats we cannot rely on this remaining true for long. Sooner or later, bacteria will evolve so that they survive well in the presence of antibiotics, without serious costs to themselves. Our only chance is therefore to use antibiotics sparingly, confining use to situations where they are really needed, and making sure that all infecting bacteria are killed quickly, before they have time to evolve resistance. If one stops treatment while some bacteria remain present, their population will inevitably include some resistant bacteria, which can then spread to other people. Antibiotic resistance can also spread between bacteria, even ones of different species. Antibiotics given to farm animals, to keep infections down and promote growth, can cause resistance to spread to human pathogens. Even these consequences are not the whole of the problem. Bacteria that have resistance mutations are not typical of their populations, but sometimes have higher mutation rates than average, allowing them to respond even faster to selection.

Drug and pesticide resistance evolve whenever drugs are used to kill parasites or pests, and literally hundreds of cases have been studied in microbes, plants, and animals. Even the HIV virus mutates within AIDS patients treated with drugs, and evolves resistance so that the treatment eventually fails. To try to prevent this, two drugs instead of one are often used. Because mutations are rare events, the virus population in a patient is unlikely to get both resistance mutations very quickly, but this usually happens eventually.

These examples illustrate natural selection, but, like artificial selection, they involve situations where the environment is changing as a result of human intervention. Many other human activities are causing evolutionary changes in organisms. For example, it seems that killing elephants for their ivory has led to increased frequencies of tuskless elephants. In the past, these were rare, abnormal animals. Now, with intensive hunting, these animals can survive and reproduce better than normal ones, and as a consequence they are increasing in elephant populations. Swallowtail butterflies with small wings, which are poor flyers, are being selected in fragmented natural habitats, presumably because individuals that do not fly far are more likely to remain within suitable habitat patches. Humans also select for an annual life-history, with speedy seed production, when we remove weeds from gardens or crop fields. In species such as the grass *Poa annua*, individuals exist that develop more slowly, and can live two years or more, but these are at a clear disadvantage in a regime of intensive weeding. These examples not only show how common and rapid evolutionary change can be, but also that anything we do may affect the evolution of species associated with humans. With people spreading all over the planet, few species will remain unaffected.

Biologists have also studied many cases of selection that are entirely natural, not involving human habitat degradation or alteration. One of the best is the 30-year study by Peter and Rosemary Grant of two species of Darwin's finch, the ground finch and the cactus finch, on the island of Daphne in the Galapagos islands (see Chapter 4). These species differ in their mean beak sizes and shapes, but there is considerable variation within each species for both characters. During the study, the Grants' team systematically ringed and measured every bird hatched on the island, and the offspring of every female were identified. Their survival through life was followed and related to measurements of size and shape of body parts. Pedigree studies showed that the variation in beak characters has a strong genetic component, so that offspring resemble their parents. Studies of the feeding behaviour of the birds in the wild show that beak size and shape affect the efficiency with which the birds deal with different types of seed: large, deep beaks allow birds to handle large seeds better than small, shallow beaks, while the reverse is true for small seeds. The Galapagos are subject to episodes of severe drought, associated with the El Niño phenomenon, and these affect the abundances of different types of food. In a drought year, most food plants fail to produce seeds, except for a species that produces very large seeds. This means that birds with large, deep beaks have a much better chance of survival than others, as was seen directly from the population censuses: after an episode of drought, the surviving adults in both species had larger, deeper beaks than the population before the drought. In addition, their offspring inherited these characters, so that the change in direction of selection caused by the drought induced a genetic change in the composition of the population – a real evolutionary change. The extent of this change agreed with that predicted from the observed relation between mortality and beak characters, taking into account the degree of resemblance between parents and offspring. Once normal conditions were restored, the relations between beak characters and survival changed in such a way that large, deep beaks were no longer favoured, and the populations evolved back towards the previous state. However, even in non-drought years there were also more minor changes in the environment that resulted in changes in the relation between fitness and beak traits, so there were constant fluctuations in beak characteristics over the whole 30 years, with the populations of both species ending up significantly different from the initial state.

Another good example is provided by the way in which flowers are adapted to their insect and other animal pollinators. For a plant to mate with others of its own species, pollinators must be attracted to visit the plant's flowers, and be rewarded for doing so (by nectar or excess pollen that they can eat), which ensures that they will visit other plants of the same species. Both the plant and animal players in these interactions evolve to get the best they can for themselves. For an orchid, for instance, it is important that a pollinating moth probes the flowers deeply, in order that the pollen mass (called a pollinium) gets firmly attached to the moth's head when it visits. This ensures that the pollinium makes good contact with the right part of the flower which the moth visits next, so that it engages correctly and the pollen will fertilize the flower. The need to keep the nectar almost out of reach of the moths' tongues drives natural selection on nectary tube length, and flowers with deviant lengths of nectary tubes should therefore have lower fertility. Flowers with shorter tubes will allow moths to suck nectar without picking up pollinia or depositing them, and flowers whose tubes are too long will waste nectar, like juice boxes whose straws are invariably too short to get all the juice out. In the juice box industry, waste benefits juice sellers, enabling them to sell larger amounts, but plants lose energy, water, and nutrients if they make nectar that is useless, and these resources could be put to better use.

In a South African *Gladiolus* species that produces only one flower per plant, individuals with longer tubes produced a fruit more often than those with average tubes, and also had more seeds per fruit than the average. This species' tubes are on average 9.3 cm long, and their hawkmoth visitors' tongues are between 3.5 to 13 cm. Moths that had no pollen on their tongue all had the longest tongues. Other hawkmoth species at the same locality, which do not pollinate this species, have tongues averaging less than 4.5 cm. This shows the power of selection to push the flowers and moths to adapt to one another, reaching extremes in some cases. Some Madagascan orchids have flowers whose nectaries are 30 cm long, and their pollinators' tongues are 25 cm. In these species, selection on length has been demonstrated by experiments in which nectar spurs were tied off to shorten them, leading to lower success in getting the moths to remove pollinia.



A similar kind of selection and counter-selection affects our own species in its relations with parasites. Several different human adaptations to malaria have been well studied, and we have clearly evolved a number of different defences, including changes in our red blood cells, where malaria parasites spend part of their complex life-cycle. Like warfarin resistance in rats, defences may sometimes have costs. The disease sickle-cell anaemia, which is usually fatal in the absence of medical treatment, involves a changed haemoglobin (the major red blood cell protein, responsible for carrying oxygen round the body). The changed form (haemoglobin S) is a variant form of the gene that codes for the common adult haemoglobin, A, and the two versions differ by a single DNA letter. Individuals whose genes for this protein are both the S type suffer from sickle-cell anaemia; their red blood cells become malformed and clog tiny blood vessels. People with one normal haemoglobin A and one S version of the gene are not affected, but have the benefit of higher resistance to malaria compared with people with two haemoglobin

A genes. The disease suffered by people with two S genes is thus a cost of malaria resistance, and prevents the S form from spreading throughout the population, even in areas with high levels of malarial infections. The variants of the enzyme glucose-6-phosphate dehydrogenase that also help protect against malaria (see Chapter 3) have a cost as well, at least when people with these variants eat certain foods or drugs, causing damage to their red blood cells, whereas the non-resistant version of this enzyme prevents this. Resistance to malaria with no, or very slight, costs does seem to be possible, however. The blood type Duffy-, another red blood cell characteristic, is widespread in much of Africa, and people with this group are much less susceptible to a certain type of malaria than people who carry the alternative Duffy+ type.

Malaria resistance illustrates a common finding, that different responses can occur to a single selective pressure, in this case a serious disease. Some of the solutions to the problem posed by malaria are better than others, because there are lower costs for the people involved. In fact, there are many other genetic variants found in different human populations that confer resistance to malaria, and it seems to be largely a matter of chance which particular type of mutation becomes established by selection in a given locality.



The examples just discussed illustrate selection responses to changes in the environment of animals, humans, and plants. Perhaps a disease appears, and there is selection on a population so that resistant individuals evolve. Or perhaps a moth evolves a longer tongue and can take nectar from flowers without picking up pollen, so that the flower in turn evolves longer nectaries. In these examples, natural selection changes the organisms, as Darwin envisaged in the 1858 quotation we gave in Chapter 2. Natural selection, however, also often acts to prevent changes. In Chapter 3, when we described the cellular machinery of proteins and enzymes, we mentioned that mutations happen and can degrade these functions. Even in a constant environment, selection acts on every generation against mutant genes (that encode mutant proteins or ones that are expressed in the wrong place or time, or in the wrong amount). New individuals with mutations arise in every generation, but non-mutants tend to leave more offspring and so their genes remain the most common, with the mutated versions remaining at low frequencies in the population. This is *stabilizing* or *purifying* selection, keeping everything working as well as possible. An example is the gene that codes for one of the proteins involved in blood clotting. Some changes to the sequence of the protein result in an inability of blood to clot following a cut (haemophilia). Until quite recently, when the causes of haemophilia were understood and it became possible to help haemophiliacs by injecting clotting factor proteins, this condition was usually lethal or severely reduced survival. Thousands of low-frequency genetic variants with such deleterious effects, affecting every conceivable characteristic, have been described by medical geneticists.

Stabilizing selection occurs if the environment has remained fairly constant, so that selection in the past has had time to adjust a trait to the state which confers high fitness. It can be detected acting today on continuously variable characters of organisms. A well-studied example is human birth weight. Even today, when very few babies die, babies with intermediate weights survive best. The low level of infant mortality largely involves tiny babies, and some very large ones. Stabilizing selection has also been observed in animal species, such as birds and insects, after severe storms, when surviving individuals tend to be intermediate in size, while the smallest and largest are often lost. Even minor deviations from an optimum may lower survival or fertility. It makes sense, therefore, that organisms' adaptation to their environment is often impressive. As we explained in Chapter 3, it sometimes seems as if even the tiniest detail can be important. Near perfection is often achieved, such as the extraordinary accuracy of cryptic butterflies' resemblance to leaves or of caterpillars to twigs. Stabilizing selection also makes sense of the observation that species often show little evolutionary change; as long as their environment poses no new challenges, selection will tend to keep things as they are. The stable morphology of some organisms over long evolutionary times, such as the so-called *living fossils* whose modern members resemble distantly related fossils, can thus be understood.

## Sexual selection

Natural selection is the only explanation for adaptation that has stood up to empirical tests. However, selection does not necessarily increase the overall survival or number of offspring produced by the population as a whole. When there is competition, traits that confer success in competition for a limited resource may reduce everyone's survival. If the most successful type of individuals become common in a population, the population's survival probability may be decreased. Examples of the maladaptive consequences of competition are not confined to biological situations. The intrusiveness and frequent bad taste of advertising are well known.



One of the best understood biological examples of competition is selection acting on the ability of males to obtain mates. In many animal species, not all fertile males leave descendants, but only those that succeed in courtship and/or in contests with other males. Sometimes, only the 'dominant' males are accepted by the females. Even fruitfly males have to court females – with dances, songs (produced by beating their wings), and scents – before being allowed to mate. Success is not guaranteed (not surprisingly, since females must be choosy and avoid accepting males of the wrong species). In many mammals, such as lions, there are social hierarchies in the ability to get matings, and females are choosy, so that males differ in their reproductive success. Natural selection will therefore favour characteristics associated with males' dominance in the hierarchy, or their attractiveness to females. Male deer have large antlers, which are used in fights between males, and some species have other means of intimidation, such as loud roaring. If

**17. The outcome of sexual selection, as illustrated in Darwin's *The Descent of Man and Selection in Relation to Sex*. The figure shows a male and female of the same species of bird of paradise, showing the male's ornamentation and the female's lack of display.**

these characters are heritable (which, as we saw above, is often the case), males with the characteristics that make them successful in mating will pass on their genes to many progeny, while other males will tend to have fewer offspring.

Both sexes may evolve characteristics by this *sexual selection*, and it probably accounts for the bright plumage of many birds. However, in many species these characteristics are confined to males (Figure 17), suggesting that they are not in themselves good adaptations to the species' environment. Many such male characters certainly do not seem likely to help survival, and they often incur costs because of lower survival of their male carriers. Peacock males, with their enormous and beautiful tails, are poor flyers, and they would probably be better able to escape predators if their tails were smaller. Peacocks are an inconvenient species for experimental studies of the aerodynamics of flight, but swallows' tails have been shown to be longer than optimal for flight, while males with longer tails are preferred by females. Even less spectacular male courtship characters often bring increased risks. For example, some tropical frog species are preyed on by bats that detect males singing their courtship songs. Even without these dangers, courting males often expend large amounts of effort, which could be otherwise employed, for example in looking for food, and they are often in extremely poor physical shape at the end of the mating season.



Realizing this, Darwin considered selection in the context of courtship to be different from most other situations, and introduced the special term sexual selection to highlight this difference. As we have just argued, it is unlikely that male peacocks' tails are adaptive, both on *a priori* grounds (such tails do not look like a good design for a flying animal), and because, if they *were* good, females should have them too. It therefore seems that selection has traded reduced flying ability against increased male mating successes in peacocks, a species in which competitive mating is important. Thus sexual selection again shows that the word fitness as used in biology often means something different from the everyday use of the word. A peacock male handicapped by his tail is not 'fit' in the meaning of being a good flyer or runner (although he may be unable to produce a fine tail if he is not well nourished and healthy), but in the shorthand of evolutionary biology he has high fitness; without his large tail, the females would mate with other males and his fertility would be low.

# Chapter 6
# The formation and divergence of species

One of the most familiar facts of biology is the division of living forms into recognizably different species. Even the most casual observation of the birds living in a north-western European town, for example, shows the presence of several species: the robin, blackbird, song thrush, missel thrush, blue tit, great tit, pigeon, sparrow, chaffinch, starling, and so on. Each species has its distinctive body size and shape, plumage coloration, song, and feeding and nesting habits. A different but broadly similar array of bird species can be found in eastern North America. Males and females of each species pair only with each other, and their offspring of course belong to the same species as their parents. Within a given geographical location, sexually reproducing animals and plants can nearly always readily be assigned to distinct groups (although careful observation sometimes reveals the existence of species with only very slight anatomical differences). Different species that coexist in the same locality remain distinct because they do not interbreed. Most biologists regard this lack of interbreeding (*reproductive isolation*) as the best criterion for defining different species. The situation is more complex with organisms that do not reproduce regularly by sexual matings, such as many kinds of microbes, and we will defer discussion of these until later.

## The nature of differences between species

Although, like the force of gravity, this division of living organisms into discrete species is so familiar that we take it for granted, it is not an obviously necessary state of affairs. It is easy to imagine a world without such sharp differences; in the bird example above, there could be creatures that combine the characteristics of, say, robins and thrushes in different proportions, and in which a mating between a given pair of parents would yield offspring with widely different character combinations. If there were no barriers to interbreeding between members of different species, the diversity of life that we see in the world could not exist, and there would be something approaching a continuum of forms. In fact, when for one reason or another barriers to interbreeding between formerly separate species have broken down, such highly variable offspring are indeed produced.

A fundamental problem for evolutionists is therefore to explain how species come to be distinct, and why reproductive isolation exists. This is the main topic of this chapter. Before embarking on it, we will describe some of the ways in which closely related species are prevented from interbreeding. Sometimes, the main barrier is a simple difference in habitats or in the time of breeding of the species. In plants, for example, there is often a characteristic brief flowering time each year, and species with non-overlapping flowering times will obviously be unable to interbreed. In animals, the use of different breeding sites may prevent individuals from different species from mating with each other. Subtle features of organisms, which can only be discovered by detailed studies of the species' natural history, often prevent individuals from different species from successfully mating with each other, even if they come together in the same place at the same time. For example, there may be an unwillingness to court individuals of the other species, because they do not produce the right smell or sound, or their courtship displays may differ. Behavioural barriers to mating are obvious in many animals, and plants have chemical means of detecting pollen from the wrong species and rejecting it. Even if mating takes place, sperm from the wrong species may be unsuccessful in fertilizing the eggs of the female.

Some species are, however, sufficiently closely related that they will occasionally mate, especially if given no choice of a member of their own species (for example, dogs, coyotes, and jackals, mentioned in Chapter 5). In many such situations, however, the first-generation hybrids often fail to develop; experimental crosses between individuals belonging to different species often produce hybrids that die at an early stage of development, whereas most offspring of crosses between individuals of the same species develop to maturity. Sometimes hybrid individuals can survive, but at a much lower frequency than non-hybrids. Even when hybrids are viable, they are often sterile, and do not produce any offspring which could pass genes on to future generations; mules (which are hybrids produced by crosses between donkeys and horses) are a famous example. Complete inviability or sterility of the hybrids obviously isolates the two species.



## The evolution of barriers to interbreeding

Although these different means of preventing interbreeding are familiar, it is a puzzle to understand how they could evolve. This is the key to understanding the origin of species. As Darwin pointed out in Chapter 9 of *The Origin of Species*, it is most unlikely that the inviability or infertility of interspecies hybrids could be the direct product of natural selection; there can be no advantage to an individual producing inviable or sterile offspring if hybridized with a different species. It would, of course, be advantageous to avoid mating with members of another species if the hybrid offspring are inviable or sterile, but it is difficult to see how there could be any such advantage in cases when the hybrids survive perfectly well. It therefore seems likely that most barriers to interbreeding between species are by-products of evolutionary changes that occurred after the populations became isolated from each other by being geographically or ecologically separated.

For example, imagine a species of Darwin's finch living on one of the Galapagos islands. Suppose that a small number of individuals manage to fly across to another island, previously unoccupied by this species, and successfully establish a new population. If such migration events are very rare, the new and the ancestral populations will evolve independently of each other. Under the processes of mutation, natural selection, and genetic drift, the genetic compositions of the two populations will diverge. These changes will be promoted by differences in the environments experienced by the populations, to which they become adapted. For example, the food plants available to a seed-eating species of bird differ from island to island, and even members of the same species of finch differ between islands in their beak sizes in ways which reflect differences in food abundance.

The tendency of populations of the same species to differ according to their geographical location, often in ways which are clearly adaptive, is called *geographical variation*. Obvious examples in the human species are the numerous minor physical differences between the races, as well as the smaller local differences in features such as skin pigmentation and stature. Such variability is found in many other species of animals and plants with wide geographical ranges. In a species that consists of a set of local populations, there is usually some migration of individuals between different locations. The amount of migration varies enormously between organisms; snails have very low migration rates, whereas organisms like birds or many flying insects are highly mobile. If migrant individuals can interbreed with members of the population in which they arrive, they will contribute their genetic makeup to this population. Migration is therefore a homogenizing force, opposing the tendency for local populations to diverge genetically by selection or genetic drift (see Chapter 2). Populations of a species will diverge more or less from each other, depending on the amount of migration, and on the evolutionary forces promoting differences between local populations. Strong selection can cause even adjacent populations to differ. For example, lead or copper mining produces soil contaminated with these metals, which are very toxic to most plants, but metal-tolerant forms have evolved on the polluted land surrounding many mines. In the absence of the metals, the tolerant plants grow poorly. Tolerant plants are therefore found only on or very close to the mines, and there is a sharp changeover to non-tolerant individuals at the boundaries.

In less extreme cases, gradual geographical changes in traits arise because migration blurs the differences caused by selection that varies geographically, in response to changes in environmental conditions. Many species of mammals living in the temperate zone of the northern hemisphere have larger body sizes in the north. Average body size changes more or less continuously from north to south, probably reflecting selection for a smaller ratio of surface area to volume in colder climates, where heat loss is a problem. Northern populations also tend to have shorter ears and limbs than southern populations, for similar reasons.



Differences between geographically separate populations of the same species do not necessarily require different types of selection. The same selection can sometimes lead to different responses. For example, as we described in Chapter 5, human populations in regions subject to malaria infections have different genetic mutations that confer resistance to malaria. There are multiple molecular paths to resistance. Different mutations that can cause resistance will occur by chance in different places, and it is largely luck which mutation comes to predominate in a given population. Differences between populations of the same species can also evolve even if there is no selection at all, as a result of the random process of genetic drift mentioned earlier. In many species, there are often marked genetic differences among different populations even for variants in DNA or protein sequences that have no effect on visible traits, and the human population is no exception to this. Even within Britain, there are differences in the frequencies of individuals with the A, B, and O blood groups, which are determined by variant forms of a single gene. The O blood group is more frequent in North Wales and Scotland than in the south of England, for example. Over wider areas, there are much greater differences in blood group frequencies. Blood group B has a frequency of over 30% in some parts of India, whereas it is virtually absent from native Americans.

There are many other such examples of geographical variation. Despite the visible differences between the major races, humans have no biological barriers to interbreeding between different populations or racial groups. In some species, however, populations from the extreme ends of a species range look different enough that they might well be regarded as different species, except for the fact that they are connected by a set of intergrading populations which interbreed with each other. There are even cases in which populations at opposite ends of a species range have diverged so much that they cannot interbreed; if the intermediates were to become extinct, they would constitute different species.

This illustrates an important point: on the theory of evolution, there must be intermediate stages in the development of reproductive isolation, and so we ought to observe at least some cases in which it is difficult to say whether or not a given pair of related populations belong to the same or different species. While this is inconvenient if we want to put things into cut-and-dried categories, it is a predictable outcome of evolution, and is evident in the natural world. There are many known examples of intermediate stages in the evolution of complete inability to interbreed between geographically separated populations. A particularly well-studied example is the American species of fruitfly, *Drosophila pseudoobscura*. This lives on the west coast of North and Central America, more or less continuously from Canada to Guatemala, but there is also an isolated population living near Bogotá, in Colombia. Flies from the Bogotá population look identical to those from other populations of the species, but their DNA sequences differ slightly. Since accumulation of sequence differences requires a long time, the Bogotá population was probably founded by a few migrant flies around 200,000 years ago. In the laboratory, flies from Bogotá will readily mate with *Drosophila pseudoobscura* flies from other populations; first-generation hybrid females are fully fertile, but hybrid males from the cross with non-Bogotá females as their mothers are sterile. No hybrid male sterility is observed in crosses among very different populations from the rest of the species range. If flies from the main population were introduced into Bogotá, they would presumably interbreed fairly freely with the Bogotá flies, and since the female hybrids are fertile, interbreeding could continue every generation. Thus the Bogotá population owes its distinctness purely to its geographic isolation. There is therefore no compelling reason to regard it as a separate species, although it is starting to develop reproductive isolation, as indicated by the sterility of the hybrid males.



It is relatively simple to understand why populations of the same species living in different places may come to diverge with respect to characteristics that adapt them to differences in the environment, as in the Galapagos finch example. It is less obvious why this leads to failure to interbreed. This may sometimes be a fairly direct by-product of adaptations to different environments. For example, two species of monkeyflower plants, *Mimulus lewisii* and *M. cardinalis*, grow in the mountains of the north-western USA. Like most monkeyflowers, *M. lewisii* is pollinated by bees, and its flowers show several adaptations for bee pollination (see the table on page 97). Unusually for a monkeyflower, *M. cardinalis* is pollinated by hummingbirds, and its flowers differ in several characteristics that promote pollination by hummingbirds. *M. cardinalis* thus probably evolved from a bee-pollinated ancestor, similar in appearance to *M. lewisii*, by a process of changing these flower characteristics.

Floral characteristics of two *Mimulus* species

| Species | *M. lewisii* | *M. cardinalis* |
|---|---|---|
| Pollinators | bee | hummingbird |
| Flower size | small | large |
| Flower shape | wide, with 'landing platform' | narrow, tubular |
| Flower colour | pink | red |
| Nectar | moderate, high sugar | abundant, low sugar |

The two monkeyflower species can be crossed experimentally, and the hybrids are healthy and fertile, yet in nature the species grow side by side without intermingling. Observations on pollinator behaviour in the wild show that, after visiting *M. lewisii*, bees rarely visit *M. cardinalis*, and a hummingbird that has visited *M. cardinalis* will rarely go on to a *M. lewisii* plant. To find out how pollinators would react to plants with intermediate flower traits, an artificially produced second-generation hybrid population, with a wide range of combinations of traits from the two parents, was planted in the wild. The trait that most strongly promoted isolation was flower colour, with red deterring bees and attracting visits by humming birds. Other traits affected one or the other of the two pollinators. A higher nectar volume per flower increased hummingbird visits, whereas flowers with larger petals were visited more often by bees. Intermediate forms between the two species had intermediate probabilities of being pollinated by bees versus hummingbirds, and hence intermediate degrees of isolation from the parent species. In this example, changes driven by natural selection as hummingbird pollination evolved have led to the *M. cardinalis* population becoming reproductively isolated from a closely related *M. lewisii* population.

Even though in most cases we do not know what force drove the divergence between closely related species and resulted in their reproductive isolation, the origin of reproductive isolation between a pair of geographically separated populations is not particularly surprising, if there have been independent evolutionary changes in two populations. Each alteration in the genetic composition of one population must either be favoured by selection in the population, or else must have such a slight effect on fitness that it can spread by genetic drift (discussed in Chapter 2, and at the end of this chapter). If a variant is spreading in a population because it has an advantage in adapting the population to its local environment, its spread will not be impeded by any harmful effects when combined (in hybrids) with genes from a different population which it never naturally encounters. There is no selection to maintain compatibility of mating behaviour between individuals from geographically or ecologically separated populations, or to maintain harmonious interactions that allow normal development, between genes that have come to differ in different populations. Like other characteristics that are not subject to selection to maintain them (such as the eyes of cave-dwelling animals), the ability to interbreed degenerates over time.



Given enough evolutionary divergence, complete reproductive isolation seems inevitable. It is no more surprising than the fact that electrical plugs of British design do not function in Continental European sockets, even though each type of plug functions perfectly with its own sockets. In human-designed machines where compatibility is desirable, constant efforts must be made to preserve it, for example in software for PC versus Macintosh computers. Genetic analyses of interspecies crosses show that different species really do contain different sets of genes which are dysfunctional when brought together in hybrids. As already mentioned, the first-generation male hybrids between many species of animals are sterile, while the females are fertile. Crosses are then possible between fertile hybrid females and either of the parental species. By testing the fertility of the male offspring of such crosses, we can study the genetic basis of the hybrid male sterility. This kind of study has been intensively carried out using *Drosophila* species; the results show clearly that the hybrid sterility is produced by interactions between different genes from the two species. In the case of the mainland versus Bogotá populations of *D. pseudoobscura*, for example, about 15 distinct genes which differ between the two populations seem to be involved in causing the sterility of hybrid males.

The time needed to produce sufficient differences between a pair of populations to make them incapable of interbreeding is very variable. In the *Drosophila pseudoobscura* example, 200,000 years (over a million generations) has produced only very incomplete isolation. In other cases, there is evidence for the very rapid evolution of barriers to interbreeding, as in the case of fish species of the cichlid family in Lake Victoria. Here, there are over 500 species apparently derived from one ancestral species, yet geological evidence shows that the lake has existed for only 14,600 years. Isolation between these species seems to be largely due to behavioural traits and coloration differences, and there is very little differentiation between the species in their DNA sequences. It seems to have taken around 1,000 years on average for a new species of this group to be produced, but other groups of fishes in the same lake have not evolved new species at such a high rate; typically, several tens of thousands of years seem to be needed for a new species to be formed.

Once two related populations have become completely isolated from each other by one or more barriers to interbreeding, their evolutionary fates will forever be independent of one other, and they will tend to diverge over time. One important cause of such divergence is natural selection; closely related species often differ in many structural and behavioural characteristics that adapt them to their different ways of life, as we have already described with the Galapagos finches. Sometimes, however, there are very few evident differences between related species. This is often the case with insects; for example, the *Drosophila* species *D. simulans* and

*D. mauritiana* both have very similar bodily structure, and differ externally only in the structure of the male genitalia. Nevertheless, they are true species, and are very reluctant to mate with each other. Similarly, it has recently been discovered that the common European pipistrelle bat is divided into two different species. They do not interbreed in nature, and differ in their calls as well as in their DNA sequences. Conversely, as we have already described, there are many examples of marked differences between populations of the same species, with no barriers to interbreeding.

These examples show that there is no absolute relationship between differences in easily observable characteristics and the strength of reproductive isolation between a pair of populations. Nor is the extent of differences between a pair of species very closely related to the time since they became reproductively isolated. This is illustrated by the striking differences among island species such as the Galapagos finches, which have evolved over a very short time-span compared with the amount of time that separates related South American species of birds, many of which differ far less (see Figure 13, Chapter 4). Similarly, in the fossil record, there are many examples of lineages showing little or no change over thousands or millions of years, followed by abrupt transitions to new forms, usually recognized as new species by palaeontologists.

Theoretical models, as well as laboratory experiments, show that intense selection can produce profound changes in a trait over 100 generations or less (see Chapter 5). For example, a population of the fruitfly *Drosophila melanogaster* has been selected artificially for an increase in the number of bristles on the flies' abdomens. Selection has produced a three-fold increase in average bristle number over 80 generations. This is about the same as the increase in the average size of the brain case between our earliest ape-like ancestors and ourselves, which took about 4 million years (roughly 200,000 generations). Conversely, traits will not change greatly, once a species living in a stable environment has had time to adapt to it. It is usually impossible to tell from the fossil record whether an observed 'sudden' evolutionary change implies the origin of a new species (which cannot interbreed with its progenitor), or simply involves a single lineage, evolving in response to environmental changes. In either case, there is no mystery in geologically rapid change.

Finally, what do species mean when there is asexual reproduction, which occurs in many single-celled organisms like bacteria? Here, the criterion of interbreeding is meaningless. For purposes of classification in these cases biologists simply use arbitrary measures of similarity, either based on characters of practical importance (such as the composition of bacterial cell walls), or increasingly on DNA sequence differences. Sufficiently similar individuals, which cluster together in regard to the characters that are used, are classed as the same species, whereas other groups of individuals that form a different cluster are assigned to different species.

## Molecular evolution and divergence between species

Given the erratic relationship between the time since separation of a pair of species and their divergence in morphological characteristics, biologists are increasingly using information from the DNA sequences of different species to make inferences about their relationships.

Rather like comparisons of the spelling of the same word in different but related languages, we can see similarities as well as differences in the sequences of the same genes in different species. For example, *house* in English, *haus* in German, *huis* in Dutch, and *hus* in Danish all have the same meaning, and are pronounced very similarly. There are two types of differences between these words. First of all, there are changes of letter in a given position, as in the substitution of *a* for *o* in the second position between German and English. Second, there are additions and deletions of letters; the final *e* in English is missing from the other languages, and Danish lacks the *a* in the second place in German. Without more information on the historical relationships between the languages, it is difficult to be certain about the direction of these changes, although the fact that only English has a final *e* suggests quite strongly that this is a late addition, and the fact that *hus* is the shortest version suggests that a vowel has been lost from the Danish word. Given such comparisons of a large sample of words, the differences between different languages can be used to measure their relationships, and the differences correlate well with the time the languages have been diverging. American English is only a couple of hundred years separated from British English, but has diverged quite noticeably, including the development of different local versions. Dutch and German are more diverged, French and Italian even more so.

The same principle can be used for DNA sequences. In this case, changes due to insertions and deletions of individual letters in the DNA are rare in the portions of genes that code for proteins, since these will usually have major effects on the sequence of amino acids in the protein and render it non-functional. Between closely related species, most changes in the coding sequences of genes involve single changes to individual letters of the DNA sequence, such as changing a G to an A. An example is given in Figure 8, which shows sequences of portions of the melanocyte-stimulating hormone receptor gene from humans, chimpanzees, dogs, mice, and pigs.

By comparing the numbers of letters in the DNA by which the sequences of the same gene differ between a pair of different organisms, one can quantify their level of divergence precisely, which is difficult to do with morphological similarities and differences. Knowing the genetic code, we can see which of the differences alter the protein sequence corresponding to the gene in question (*replacement* changes), and which do not (*silent* changes). For instance, in the melanocyte-stimulating hormone receptor sequences, a simple count of the differences between the human and chimpanzee sequences in Figure 8 reveals four differences in the 120 DNA letters shown. For the entire sequences of the different species (omitting a small region with some additions and deletions of DNA letters), the numbers of differences from the human sequence are shown in the table below.

| Human versus | Same amino acid (silent differences) | Different amino acid |
|---|---|---|
| Chimpanzee | 17 | 9 |
| Dog | 134 | 53 |
| Mouse | 169 | 63 |
| Pig | 107 | 56 |

A recent study showed that the sequence divergence for 53 non-coding DNA sequences compared between humans and chimpanzees ranged between 0 and 2.6% of the total letters, and averaged only 1.24% (1.62% for the human and gorilla). These estimates show why it is now accepted that chimpanzees, rather than gorillas, are our closest living relatives. The differences are much greater if humans are compared with the orang-utan, and greater still if we are compared with baboons. More distantly related mammals, such as carnivores and rodents, differ at the sequence level much more than do different primates; mammals differ much more from birds than they do from each other, and so on. The patterns of relationships revealed by sequence comparisons are in broad agreement with what is expected from the times at which the major groups of animals and plants appear in the fossil record, as expected in the theory of evolution.

The table of sequence differences shows that silent changes are generally much more common than replacement changes, although even silent changes are rare between the most closely related species such as chimpanzees and humans. The obvious interpretation is that most changes to the amino acid sequence of a protein impair its function to some extent. As we described in Chapter 5, a small detrimental effect caused by a mutation will result in selection quickly eliminating the mutation from the population. Most mutations that change protein sequences therefore never contribute to evolutionary differences in gene sequences that accumulate between species. But there is also increasingly firm evidence that some amino acid sequence evolution is driven by selection acting on occasional favourable mutations, so that molecular adaptation occurs (see Chapter 5).

In contrast to the often detrimental effects of mutations changing amino acids, silent changes to the sequences of genes will have little or no effect on biological functions. It thus makes sense that most divergence in gene sequences between species are silent changes. But when a new silent mutation appears in a population, it is just a single copy among thousands or millions of copies of the gene in question (two in each individual in the population). How does such a mutation spread through the population if it does not confer any selective advantage to its carrier? The answer is that random changes in the frequencies of alternative variants (genetic drift) take place in finite populations, a concept we introduced briefly in Chapter 2.



This process works as follows. Suppose that we study a population of the fruitfly *Drosophila melanogaster*. For the population to be maintained, each adult must contribute on average two descendants to the next generation. Suppose that the population varies in eye colour, with some individuals carrying a mutant gene that makes the eyes bright red while the non-mutant version of this gene makes all the other flies' eyes the normal dull red. If individuals with either type of gene have the same average number of offspring, there is no selection on eye colour; it is said to be *neutral* in its effects. Because of this neutrality with respect to selection, the genes of the next generation will be drawn randomly from the parental population (Figure 18). Some individuals may have no offspring, while others may happen by chance to have more than the average of two offspring. This means that the frequency of the mutant gene in the progeny generation will not be the same as its frequency among the parents, since it is extremely unlikely that individuals with and without the mutant gene contribute exactly the same numbers of offspring. Over the generations, there will thus be continual random fluctuations in the composition of the population, until sooner or later either all members of the population have the gene for bright red eyes, or else it is lost from the population and they all have the alternative version of the gene. In a small population, genetic drift is fast, and it will not take long until all members of the population become the same. This will take much longer in a large population.

This illustrates two effects of genetic drift. First, while a new variant is drifting to eventual loss or to a frequency of 100% (*fixation*), the character affected by the gene is variable within the population. The input of new neutral variants by mutation and the changes in variant frequencies (and, from time to time, loss of variant genes) by drift determines the variability in the population. Examination of DNA sequences of the same gene from different individuals from a population reveals variability at silent sites due to this process, as we mentioned in Chapter 5.

A second effect of genetic drift is that a selectively neutral variant that is initially very rare has some chance of spreading throughout the whole population and replacing alternative variants, although it has a much greater chance of being lost. Genetic drift thus leads to evolutionary divergence between isolated populations, even without any selection promoting the changes. This is a very slow process. Its rate depends on the rate at which new neutral mutations arise, as well as the rate at which genetic drift leads to replacement of one version of a gene by a new one. Remarkably, it turns out that the rate of DNA sequence divergence between a pair of species depends only on the rate of mutation per DNA letter (the frequency with which a particular letter in a parent is mutant in the copy that is passed to an offspring). An intuitive explanation for this is that, if

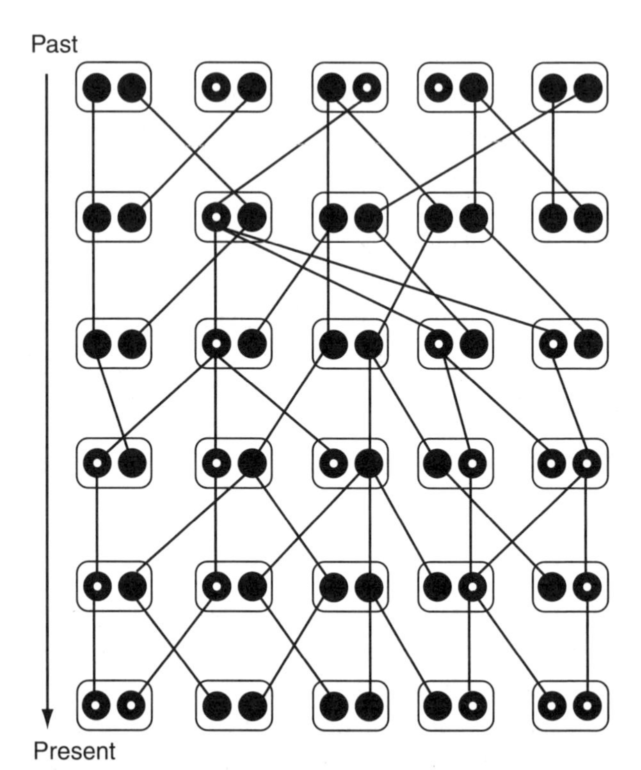


**18. Genetic drift. The process of genetic drift of a single gene over six generations, in a population of five individuals. Each individual (symbolized by an open shape) has two copies of the gene, one from each parent. The different DNA sequences of the individuals' gene copies are not shown in detail, but are symbolized by black discs with or without a white spot. The white spots might correspond to the variant gene causing bright red eye colour, and the black discs to the variant with dull red eye colour, in the *Drosophila* example given in the text. In the first generation, three individuals have one of the white spot type of the gene and one of the black type. Thus 30% of the genes in the population have the white spot. The figure shows the lines of descent of the genes in each generation (we assume for convenience that individuals can reproduce as either male or female, as is true for many hermaphroditic species of plants, such as tomatoes, and some animals, such as earthworms). Some individuals happen by chance to have more offspring than others, while other have less, or may even leave no surviving descendants (e.g. the individual shown at the right in generation 2). The numbers of white spot and black gene copies therefore fluctuate from each generation to the next. In the third generation, three individuals all inherit the white spot gene copy from the single individual carrying one such gene in generation 2, so this type of gene goes from 10% to 30%; in the next generation it is 50%, and so on.**

no selection is acting, nothing affects the number of mutational differences between a pair of species except the rate at which mutations appear in the sequence and the amount of time since the species' last common ancestor. A large population has more new mutations per generation, simply because there are more individuals in which a mutation might happen. But genetic drift happens faster in a small population, as explained above. It turns out that the two opposing effects of population size cancel out exactly, and so the mutation rate determines the rate of divergence.

This theoretical result has important implications for our ability to determine the relationships between different species. It implies that neutral changes accumulate in a gene as time goes on, at a rate that depends on the gene's mutation rate (the molecular clock principle, which we mentioned, but did not explain, in Chapter 3). Sequence changes in genes are therefore likely to take place in a much more clock-like fashion than changes in traits subject to selection. Rates of morphological changes depend strongly on environmental changes, and variable rates and reversals of direction can occur.

Even the molecular clock is not very precise. Rates of molecular evolution can change over time within the same lineage, as well as between different lineages. Nevertheless, use of the molecular clock allows biologists to roughly date the divergence between species for which there is no fossil evidence. To calibrate the clock, one needs sequences from the closest available species whose divergence dates are known. One of the most important applications of this method has been to date the timing of the split between the lineage giving rise to modern humans and the one leading to chimpanzees and gorillas, for which no independent fossil evidence is available. Use of the molecular clock with a large number of gene sequences has enabled a date of 6 or 7 million years to be estimated with considerable confidence. Because the rate of neutral sequence evolution depends on the mutation rate, the clock is exceedingly slow, since the rate at which single letters in the DNA change by

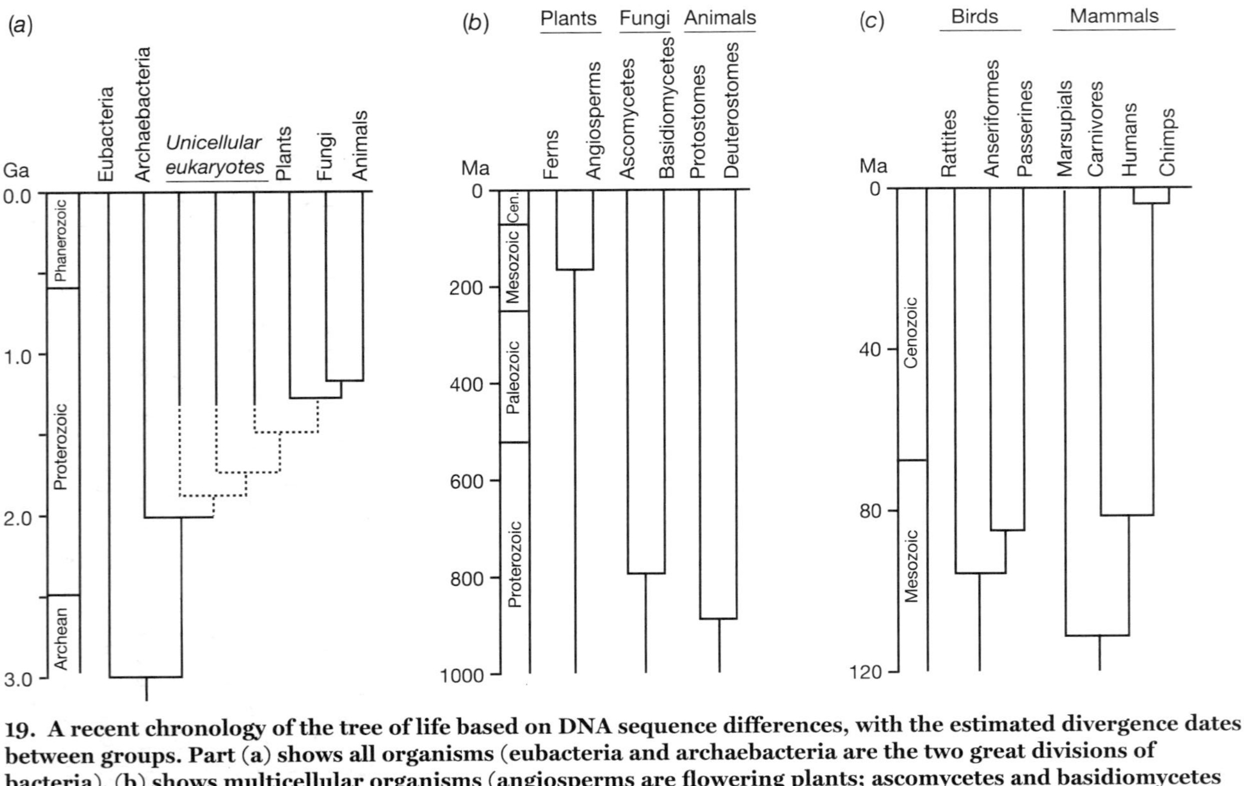


**19. A recent chronology of the tree of life based on DNA sequence differences, with the estimated divergence dates between groups. Part (a) shows all organisms (eubacteria and archaebacteria are the two great divisions of bacteria), (b) shows multicellular organisms (angiosperms are flowering plants; ascomycetes and basidiomycetes are the two major types of fungi), and (c) shows bird and mammal groups (rattites are ostriches and their relatives; anseriformes are ducks and their relatives; and passerines are the song-birds).**

mutation is very low. The fact that approximately 1% of the DNA letters differ between humans and chimpanzees corresponds to a single letter changing only once in over a billion years. This is consistent with experimental measurements of mutation rates.

A molecular clock is also found to apply to the amino acid sequence of proteins. As already mentioned, protein sequences evolve more slowly than silent DNA differences, and are therefore useful for the difficult task of comparing species that diverged a very long time ago. Between such species, multiple changes will have occurred at some sites in their DNA sequences, so that it becomes impossible to count accurately the number of mutations that have happened. Scientists who are interested in reconstructing the times of divergence between the major groups of living forms therefore use data from slowly evolving molecules (Figure 19). Such dates are, of course, rough estimates, but the accumulation of estimates from many different genes can improve the accuracy of the procedure. Judicious use of sequence information from genes that evolve at different rates is allowing evolutionary biologists to form a picture of the relationships between groups of organisms whose last common ancestors lived a billion or more years ago. In other words, we are getting close to reconstructing the genealogical tree of life.

# Chapter 7
# Some difficult problems

As the theory of evolution has become increasingly well understood and tested by biologists, new questions have arisen. Not all problems have been solved, and there is still debate about old questions as well as new ones. In this chapter, we describe some examples of biological phenomena that are apparently difficult to explain. Some of these were dealt with by Darwin himself, others have been the subject of later research.

## How can complex adaptations evolve?

Critics of the theory of evolution by natural selection frequently raise the difficulty of evolving complex biological structures, from protein molecules through single cells to eyes and brains. How can a fully functioning and beautifully adapted piece of biological machinery be produced purely by selection acting on chance mutations? The key to understanding how this can happen is expressed in another meaning of the word 'adapt'. In the evolution of organisms and their complex machinery, many aspects are modified (adapted) versions of pre-existing structures, just as when machines are made by engineers. In making complex machines and devices, less elegant initial models are refined over the course of time and diversified (adapted) to new, sometimes unanticipated, uses. The evolution of the total knee replacement is a good example of the process by which a crude initial solution to a problem was good enough to be useful, but was successively adapted to work better and better. Just as in biological evolution, many early designs were developed that seem poor by today's standards, yet each was an improvement on the ones before, and could be used by knee surgeons. These each played their roles as stages in the evolution of modern, complex artificial knees.

This process of successive adaptation of 'designs' is like climbing a hill in a thick fog. Even without a goal of reaching the top (or even without knowing where it is), if one follows a simple rule – each step goes uphill – one will move closer and closer to the summit (or at least to a local top). Simply by making a structure work better in one way or another, the end result is an improved design, without a Designer being necessary. In engineering, improved design is often the result of many contributions from different engineers over the evolution of a machine, and early car designers would have been astonished at modern cars. In natural evolution, it results from what has been called 'tinkering' with the organism, with minor changes that make their possessors survive or reproduce better than others. In the evolution of a complex structure, several different traits must, of course, evolve simultaneously, so that the different parts of the structure are well adapted to function as a whole. We saw in Chapter 5 that advantageous traits can spread through a population over a short time, relative to the time available for major evolutionary changes, even if they are initially very rare. A succession of small changes to a structure that already works, but can be improved, can therefore produce large evolutionary changes. After many thousands of years, the radical transformation of even a complex structure is not difficult to imagine. After enough time, the structure will differ from its ancestral state in many different ways, so that individuals in the descendant population would have combinations of characteristics never seen in the ancestral population, just as modern cars have many differences from early cars. This is not just a theoretical possibility: as we described in Chapter 5, animal and plant breeders routinely accomplish this by artificial selection. There is thus no difficulty in seeing how natural selection can cause the evolution of highly complex characters, made up of numerous mutually adjusted components.

The evolution of protein molecules is sometimes posed as an especially difficult problem. Proteins are complex structures whose parts must interact to function properly (many proteins must also interact with other proteins and other molecules, including DNA in some cases). The theory of evolution must certainly be capable of accounting for protein evolution. There are 20 different kinds of amino acids, so the chance that the right one would appear at a particular site in a protein molecule 100 amino acids long (shorter than many real proteins) is 1 in 20. The chance is evidently vanishingly small that, if 100 amino acids were randomly thrown together, each position in the sequence would have the right amino acid, and a working protein would form. It has therefore been claimed that the chance of assembling a functioning protein is similar to that of an airliner being assembled by a tornado blowing through a scrapyard. It is true that a functioning protein could not be assembled by randomly picking an amino acid for each position in the sequence. But, as the explanation given above makes clear, natural selection does not work like this. Proteins probably started as short chains of a few amino acids that could cause reactions to go a bit faster, and were successively improved as they evolved. There is no need to worry about the many millions of potential non-functional sequences that will never exist, provided that protein sequences during evolution started off catalysing reactions better than when no protein is present, and then got successively better over evolutionary time. It is easy to see in principle how successive stepwise changes, each one changing the sequence or adding to its length, could improve a protein.

Our knowledge about how proteins function supports this. The part of a protein that is essential for its chemical activity is often only a very small part of its sequence. A typical enzyme has just a handful of amino acids that physically interact with the chemical that is to be changed by the enzyme. Most of the rest of the protein chain simply provides a scaffold that supports the structure of the part involved in this interaction. This implies that the functioning of a protein depends critically on only a relatively small set of amino acids, so that a new function could evolve by a small number of changes to the sequence of the protein. This has been verified by numerous experiments in which artificially induced changes to protein sequences have been subjected to selection for new activities. It has proved surprisingly easy to produce quite radical shifts in the biological activity of proteins by these means, sometimes just by a change in a single amino acid, and there are similar examples among naturally evolved changes.

A similar answer can be given to the question of how it is possible for pathways of successive enzyme reactions to evolve, such as those which make chemicals that organisms need (see Chapter 3). One might think that, even if the end-products are useful, it would be impossible to evolve such pathways, since evolution has no foresight and cannot build up a chain of enzyme reactions until its function is complete. Again, the solution to this apparent riddle is simple. Many useful chemicals were probably present in the environment of early organisms. As life evolved, these would become scarce. An organism that could change a similar chemical into the useful one would benefit, and so an enzyme could evolve to catalyse that change. The useful chemical would now be synthesized from the related one. Thus a short biosynthetic pathway, with a precursor and a product, would be favoured. By successive steps like this, pathways could evolve – backwards from their end-products – to build up the chemicals organisms need.

If complex adaptations really evolve in steps, as evolutionary biologists propose, we should be able find evidence for intermediate stages in the evolution of such characters. There are two sources of such evidence: the existence of intermediates in the fossil record, and present-day species that show intermediate stages between simple and more advanced states. In Chapter 4, we described examples of intermediate fossils linking very different forms; these support the principle of stepwise evolutionary changes. Of course, in many cases there is a complete absence of intermediates, especially as we go further back in time. In particular, the major divisions of multicellular animals, including molluscs, arthropods, and vertebrates, nearly all appeared rather suddenly in the Cambrian (more than 500 million years ago), with virtually no fossil evidence concerning their ancestors. Recent DNA sequence studies of the relationships between them suggests strongly that these groups were already separate lineages long before the Cambrian era (Figure 19), but we simply have no information on what they looked like, probably because they were soft-bodied and hence unlikely to fossilize. But the incompleteness of the fossil record does not mean that intermediates did not exist. New intermediates are constantly being discovered. A recent one is a 125-million-year-old mammalian fossil from China with features similar to those of modern placental mammals, but more than 40 million years older than the oldest previously known fossil of this kind.



The other type of evidence, from comparisons of living forms, is our only source of information on features that do not fossilize. A simple but compelling example is provided by flight, as pointed out by Darwin in Chapter 6 of *The Origin of Species*. There are no fossils connecting bats with other mammals; the first bat fossils, found in deposits over 60 million years old, have the same highly modified limbs as modern bats. But there are several examples of modern mammals which have the ability to glide but cannot fly. The most familiar are the 'flying' squirrels, which are very similar to ordinary squirrels except for flaps of skin connecting their fore- and hindlimbs. These act as a crude wing, which allows the squirrels to glide some distance if they launch themselves into the air. Similar adaptations to gliding have evolved independently in other mammals, including the so-called flying lemurs (which are not true lemurs, and are not related to flying squirrels), and in the marsupial sugar-gliders. Gliding species of lizards, snakes, and frogs are also known. It is easy to imagine how being able to glide reduces the risk of a tree-living animal being caught and eaten by a predator, and that gliding could evolve by a gradual modification of the body of an animal that jumps from branch to branch. A gradual increase in the area of skin used for gliding, and modifications of the forelimbs to support such an increase, would clearly be advantageous. The flying lemur has a large extensible membrane stretching from head to tail. This is very close to the wings of bats, although the animals can only glide, not fly. Once a wing structure that allows highly efficient gliding has evolved, the development of wing musculature to produce power strokes can readily be envisaged.

The evolution of eyes is another example, also considered by Darwin. The vertebrate eye is a highly complex structure, with its light-sensitive cells in the retina, the transparent cornea and lens that allow the image to be focused on the retina, and muscles that adjust the focus. All vertebrate animals have essentially the same design of eye, but with many variations of detail adapted to different modes of life. How could such a complex piece of machinery evolve, when a lens is apparently useless without without a retina, and vice versa? The answer is that a retina is certainly not useless without a lens. Many types of invertebrate animals have simple eyes, with no lens. Such animals do not need to see clearly. It is enough to perceive light and dark in order to detect predators. In fact, a whole series of intermediates between simple light-sensitive receptors and various types of complex devices that produce images of the world can be seen in different groups of animals (Figure 20). Even single-celled eukaryotes are capable of detecting and responding to light, by means of receptors composed of a cluster of molecules of the light-sensitive protein rhodopsin. Rhodopsin is used in all animal eyes, and is also found in bacteria. Starting with this simple ability of cells to detect light, it is easy to imagine a series of steps in which increased light-capturing abilities evolve step by step, leading

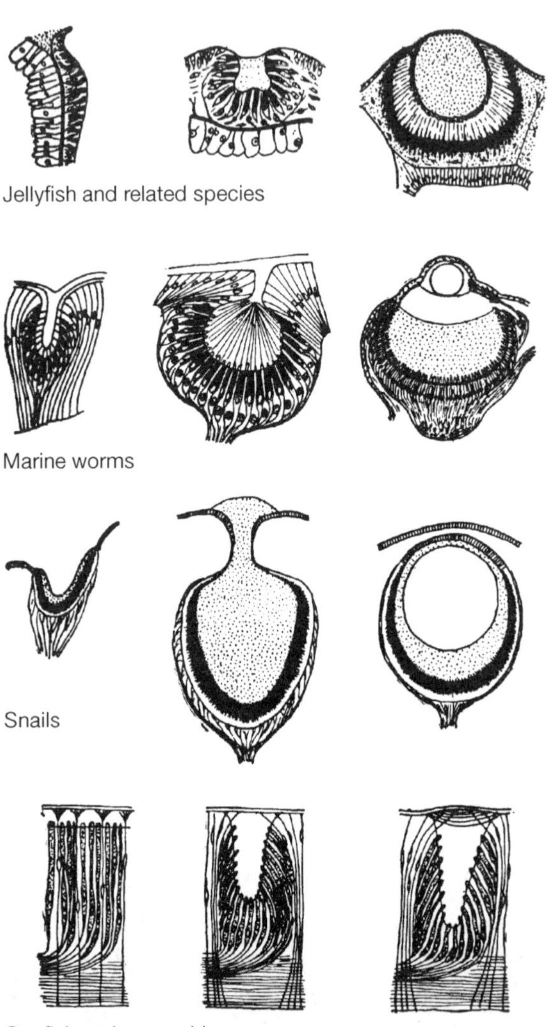


**20. Eyes of a variety of invertebrate animals. From left to right, each row shows successively more advanced types of eyes, possessed by different species within a given group. For example, in the marine worms (second row), the left-hand eye consists simply of a few light-sensitive and pigment cells with a transparent cone projecting into their midst. The middle eye has a chamber filled with transparent jelly and a retina with a large number of light-sensitive cells. The right-hand eye has a spherical lens in front of the chamber, and many more light receptors.**

eventually to a focusable lens that produces a sharp image. As Darwin put it:

> In living bodies, variation will cause the slight alterations, . . . and natural selection will pick out with unerring skill each improvement. Let this process go on for millions of years; and during each year on millions of individuals of many kinds; and may we not believe that a living optical instrument might thus be formed . . . superior to one of glass?

## Why do we age?

The bodies of young adults as a whole strike us, like the eye, as near-perfect pieces of biological machinery. The opposite problem to explaining this near-perfection is presented by the fact that it is not maintained for very long during life. Why does evolution allow this to happen? The decline of a near-perfect creature into a feeble shadow of itself as a result of ageing has been a favourite topic of the poets, especially when they foresee it happening to their lovers:

> Then of thy beauty do I question make,
> That thou among the wastes of time must go,
> Since sweets and beauties do themselves forsake,
> And die as fast as they see others grow;
> And nothing 'gainst Time's scythe can make defence
> Save breed to brave him when he takes thee hence.
>
> From 'Sonnet 12' by William Shakespeare

Ageing is, of course, not confined to humans; it has been observed in virtually every plant or animal. To measure ageing, we can study many individuals kept in a protected environment, where 'external' causes of mortality such as predation have been removed, so that individuals live for much longer than in nature. Following them over time, we can determine the probabilities of death at different ages. Mortality is usually high for very young individuals, even in protected conditions; it declines as juveniles become older and larger, but then increases again after adulthood. In most species that have been studied carefully, the adult death rate increases steadily with age. Mortality patterns, however, differ greatly in different species. Small, short-lived organisms such as mice have much higher mortality rates at relatively young ages than large, long-lived organisms like ourselves.

This senescent increase in mortality reflects the deterioration of multiple biological functions with advancing age: almost everything seems to get worse, from muscular strength to mental ability. The nearly universal occurrence of ageing in multicellular organisms (which seems like a kind of degeneration) may seem to be a severe difficulty for evolutionary theory – contradicting the idea that natural selection causes the evolution of adaptation. One answer to this is that adaptation is never perfect. Ageing is partly an unavoidable consequence of cumulative damage to the systems necessary for continued survival, and selection probably simply cannot prevent this. Indeed, the annual chance of failure of complicated machines, such as cars, also increases with age, quite similarly to the mortality of living organisms.



But this cannot be the whole story. Single-celled organisms like bacteria reproduce simply by dividing into daughter cells, and the lineages of cells produced by these divisions have persisted over billions of years. They do not senesce, but continually break down damaged components and replace them with new ones. They can continue to propagate indefinitely, provided that harmful mutations are removed by selection. This is also possible for artificially cultured cells of some organisms, such as fruitflies. The reproductive cell lineages of multicellular organisms are also perpetuated every generation, so why could repair processes not be maintained for the whole organism? Why do most of our body systems show some senescent decline? For example, mammals' teeth wear out with advancing age, eventually leading to death from starvation in nature. This is not inevitable; reptiles' teeth are renewed from time to time. The different rates of ageing of different species reflects different effectiveness of repair processes and the extent to which these are maintained with advancing age: a mouse can expect to live for about three years at best, whereas a human can live for more than 80 years. These species differences indicates that ageing evolves. Ageing therefore demands an evolutionary explanation.

We saw in Chapter 5 that natural selection on multicellular organisms works in terms of differences in individuals' contributions to the next generation, through differences in the numbers of offspring they produce, as well as in their chances of survival. Furthermore, all individuals have some risk of death from accidents, diseases, and predation. Even if the chance of death from these causes is age-independent, the chance of surviving goes down as age increases, in ourselves, as for cars: if the probability of survival from one year to the next is 90%, the chance of survival over five years is 60%, but over 50 years it is only 0.5%. Selection therefore favours survival and reproduction early rather than late in life, simply because, on average, more individuals will be alive to experience the good effects. The greater the mortality due to accidents, diseases, and predation, the more strongly will selection favour improvements early in life, relative to later on, since few individuals can survive to late ages if the death rate from these external causes is high.

This argument suggests that ageing evolves because of the greater selective value of variants with favourable effects on survival or fertility early in life, compared with variants that act late. The concept is similar to the familiar idea of life insurance: it costs less to buy a given amount of insurance if you are young, because you most likely have many years of paying ahead. There are two main ways in which natural selection might work to cause ageing. The argument used above shows that mutations with harmful effects will be most strongly selected against if they express their effects early in life. The first way that selection can cause ageing is to keep early-acting mutations rare in populations, while allowing ones with effects late in life to become common. Indeed, many common human genetic diseases are due to mutations whose harmful effects appear late in life, such as those involved in Alzheimer's disease. Second, variants that have beneficial effects early in life will be more likely to spread through the population than those whose good effects come only in old age. Improvements to the early stages of life can evolve, even if these benefits come at the expense of harmful side-effects later on. For example, higher levels of some reproductive hormones may enhance women's fertility early in life, but at the risk of later breast and ovarian cancer. Experiments confirm these predictions. For example, one can keep populations of the fruitfly *Drosophila melanogaster* by breeding only from very old individuals. In a few generations, these populations evolve slower ageing, but at the expense of reduced reproductive success early in life.



The evolutionary theory of ageing predicts that species with low externally caused death rates should evolve low rates of ageing and longer life-spans, compared with species with higher external death rates. There is indeed a strong relation between body size and the rate of ageing, smaller species of animals tending to age much faster than larger ones, and to reproduce earlier. This probably reflects the greater vulnerability of many small animals to accidents and predation. Between species with similar body sizes, striking differences in rates of ageing between animals with different mortality rates in the wild often make sense when we consider their risks of predation. Many flying creatures are notable for longevity, which makes sense because flying is a good defence against many predators. A fairly small creature like a parrot can have a life-span longer than that of a human being. Bats live much longer than terrestrial mammals like mice with comparable body weights.

We ourselves may be an example of evolution of a slower rate of ageing. Our closest relatives, the chimpanzees, rarely live beyond 50 years even in captivity, and start reproducing earlier in life than humans, at an average of 11 years of age. Humans have therefore probably substantially reduced their rate of ageing since diverging from our common ancestor with apes, and postponed reproductive maturity. These changes are probably due to increased intelligence and ability to cooperate, which reduced vulnerability to external causes of death and reduced the advantage of reproducing early. A change in the relative advantages of early versus late reproduction can be detected and even measured in present-day societies. Industrialization has led to a dramatic decline in mortality rates among adults, which is evident in census data. This changes the natural selection affecting the ageing process in human populations. Consider the degenerative brain disorder Huntington's disease, which is caused by a rare mutant gene. This disease has a late age of onset (in the 30s or later). In a population with high mortality due to disease and malnutrition, few individuals survive to their 40s, and carriers of Huntington's disease have on average only slightly (9%) fewer offspring than unaffected individuals. In industrialized societies, with low mortality, people quite often have children at ages when the disease could appear, and in consequence affected people have on average 15% fewer children than unaffected individuals. If current conditions continue, selection will gradually reduce the frequencies of mutant genes with effects expressed late in reproductive life, and the survival rates of older individuals will increase. Rare genes with major effects like Huntington's disease have only a minor effect on the population as a whole, but many other diseases which are under at least partially genetic control predominantly afflict middle-aged and elderly people, including heart disease and cancer. We may expect the incidence of these genes to decline over time due to this natural selection. If the low death rates characteristic of industrialized societies persist for several centuries (a big if), there will be a slow but steady genetic change towards lower rates of ageing.



## The evolution of sterile social castes

Another problem for evolutionary theory is posed by the existence of sterile individuals in a number of types of social animals. In social

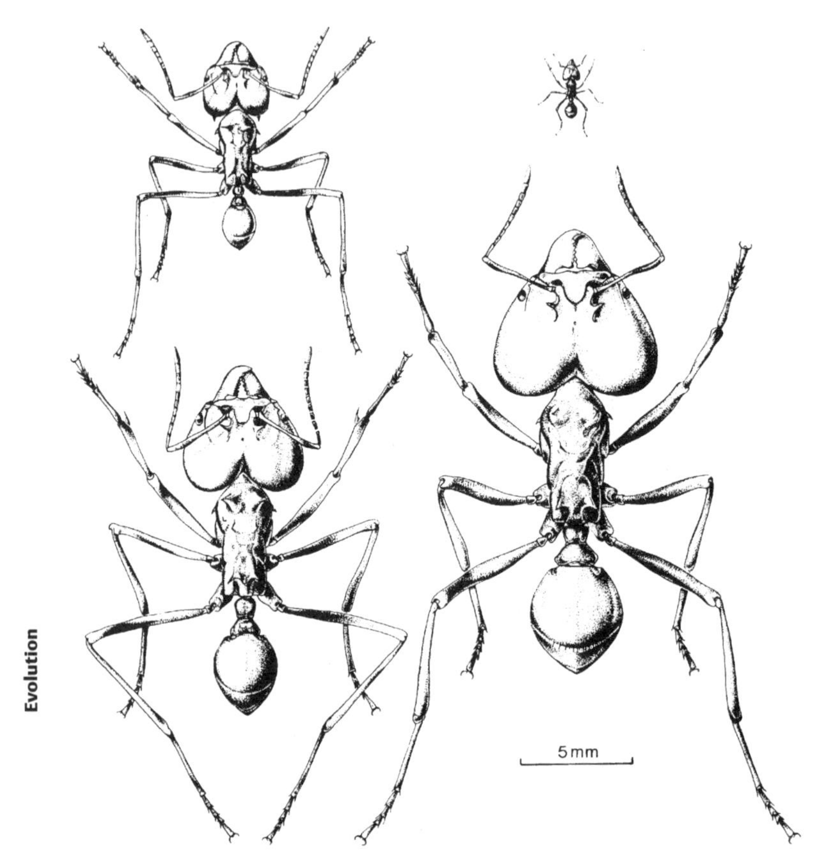


**21. Castes of workers of the leaf-cutting ant *Atta*, all from the same colony. The tiny worker at the top right tends the fungus gardens cultivated by this species. The giant individuals are soldiers, who guard the nest.**

wasps, bees, and ants, some of the females in a nest are workers, who do not reproduce. Reproductive females are a small minority within the colony (often just a single queen); the worker females look after the queens' offspring and maintain and provide for the nest. In the other main group of social insects, the termites, both males and females can behave as workers. In the advanced social insects, there are often several different 'castes', which perform very different roles and are distinguished by differences in behaviour, size, and body structure (Figure 21).

A remarkable recent discovery is that a few species of communally nesting mammals have social organizations resembling these insects, with the majority of inhabitants of a nest being sterile. The most famous is the naked mole rat, a species of burrowing rodent inhabiting desert areas of southern Africa. There may be several dozen inhabitants of a nest, with only a single reproductive female. If she dies, there is a struggle to replace her among some of the other females, in which one emerges victorious. Systems of social animals with sterile workers have thus evolved in quite different groups of animals. These species pose apparent problems for the theory of natural selection. How can individuals evolve to forego reproduction? How can the often very extreme adaptations of the castes of workers to their specialized roles have evolved, since the workers themselves do not reproduce, and so cannot be subject directly to natural selection?

These questions were raised, and partially answered, by Darwin in *The Origin of Species*. The answers lie in the fact that the members of a social group, such as a naked mole rat nest, or ant nest, are generally close relatives, often sharing the same mother and father. A genetic variant that causes its carriers to forgo their own reproductive success to help raise its relatives may help the relatives' genes pass to the next generation, and the relatives' genes are often (because of relatedness) the same as the helper individual's own genes (in the case of a brother and sister, if one individual has a genetic variant inherited from one parent, the chance is one-half that the variant will be present in the other). If the sacrifice by sterile individuals results in a sufficient increase in the numbers of surviving and reproductively successful relatives, the increase in number of copies of the 'worker gene' can outweigh the decrease due to their own lost reproductive success. The increase needed to outweigh the loss is smaller the closer the degree of relationship. J. B. S. Haldane once stated that 'I would lay down my life for two brothers or eight cousins'.

This principle of *kin selection* provides a framework for

understanding the origins of sterility in social animals, and modern research has shown that it can account for many details of animal societies, including those with less extreme features than sterile castes. For example, in some bird species, juvenile males do not attempt to mate, but remain as 'helpers' at their parental nest while younger siblings are being cared for. Similarly, wild dogs baby-sit a pack's young while other pack members go out hunting.

The question of how the differences between castes of sterile workers arise is slightly different, but has a related answer. Development as a member of a particular caste of worker is controlled by environmental cues, such as the amount and quality of food provided to the individual while a larva. However, the ability to respond to such cues is genetically determined. A certain genetic variant might confer the potentiality of a sterile member of an ant colony to develop as, say, a soldier (with bigger jaws than ordinary workers) rather than a worker. If a colony with soldiers is better defended against enemies, and if colonies with the variant can produce more reproductives on average, the variant will increase the success of its colony. If the reproductively active members of the colony are close relatives of the workers, the genetic variant that induces some workers to become soldiers will be transmitted by the colony via queens and males founding new colonies. Selection can thus act to increase the representation of this variant among colonies in the species.

These ideas also illuminate the evolution of multicellular organisms from single-celled ancestors. The cells produced from the fusion of an egg and sperm remain associated, and most of them lose the ability to become sex cells and contribute directly to the next generation. Since the cells involved are all genetically identical, this would be advantageous if survival and reproduction were sufficiently increased in the group of associated cells, compared to the single-celled alternative. The non-reproducing cells 'sacrifice' their own reproduction for the benefit of the community of cells. Some are doomed to die during the developmental process, as tissues form and dissolve, and many of them lose the potential to divide, as we explained when discussing the evolution of ageing. The serious consequences for organisms when cells regain the ability to divide without regard to the organism are manifested in cancer. The differentiation of cells into different types during development is analogous to the differentiation of castes in social insects.

## The origin of living cells and the origin of human consciousness

Two other major and largely unsolved problems in evolution, at the opposite extremes of the history of life, are the origin of the basic features of living cells and the origin of human consciousness. In contrast to the questions we have just been discussing, these are unique events in the history of life. Their uniqueness means that we cannot use comparisons among living species to make firm inferences about how they might have occurred. In addition, the lack of any fossil record for the very early history of life or for human behaviour means that we have no direct information about the sequences of events involved. This does not, of course, prevent us from making guesses about what these might have been, but such guesses cannot be tested in the ways we have described for ideas about other evolutionary problems.



In the case of the origin of life, the aim of much current research is to find conditions resembling those which prevailed early in the Earth's history, which allow the purely chemical assembly of molecules that can then replicate themselves, just as the DNA of our own cells is copied during cell division. Once such self-replicating molecules have been formed, it is easy to imagine how competition between different types of molecule could result in the evolution of more accurate and faster replicating molecules, that is natural selection would act to improve them. Chemists have been very successful in showing that the basic chemical building blocks of life (sugars, fats, amino acids, and the constituents of DNA and RNA) can be formed by subjecting solutions of simpler molecules (of the type that are likely to have been present in the oceans of the early Earth) to electric sparks and ultra-violet irradiation. There has been limited progress in showing how these can be assembled into still more complex molecules that resemble RNA or DNA, and even more limited success in getting such molecules to self-replicate, so we are still far from achieving the desired goals (but progress is constantly being made). Furthermore, once this goal is achieved, the question of how to evolve a primitive genetic code that allows a short RNA or DNA sequence to determine the sequence of a simple protein chain must be solved. There are many ideas, but as yet no definitive solutions to this problem.

Similarly, we can only make guesses about the evolution of human consciousness. It is even difficult to state the nature of the problem clearly, since consciousness is notoriously hard to define precisely. Most people would not regard a newborn baby as conscious; few would dispute that a two-year-old child is conscious. The extent to which animals are conscious is fiercely debated, but pet-lovers are well aware of the ability of dogs and cats to react to the wishes and moods of their owners. Pets even seem to be able to manipulate their owners into doing what they want. Consciousness is thus probably a matter of degree, not kind, so that in principle there is little difficulty in imagining a gradual intensification of self-awareness and ability to communicate during the evolution of our ancestors. Some would regard language ability as the strongest criterion for possession of true consciousness; even this develops gradually with age in infants, albeit with astonishing speed. Furthermore, there are clear indications of rudimentary language abilities in animals such as parrots and chimpanzees, who can be taught to communicate simple pieces of information. The gap between ourselves and higher animals is more apparent than real.

Although we know nothing of the details of the selective forces driving the evolution of human mental and language abilities, which evidently far exceed those of any other animals, there is nothing particularly mysterious in explaining them in evolutionary terms. Biologists are making rapid progress in understanding the functioning of the brain, and there is little doubt that all forms of mental activity are explicable in terms of the activities of nerve cells in the brain. These activities must be subject to control by genes that specify the development and functioning of the brain; like any other genes, these will be liable to mutation, leading to variation on which selection can act. This is no longer pure hypothesis: mutations have been found which lead to deficiencies in specific aspects of grammar in the speech of their carriers, leading to identification of a gene involved in the control of some aspects of grammar. Even the mutation in its DNA sequence that causes the difference from normal is known.

# Chapter 8
# Afterword

What have we learned about evolution in the 140 years since Darwin and Wallace first published their ideas? As we have seen, the modern view is remarkably close in many ways to theirs, with a strong consensus that natural selection is the major force guiding the evolution of structures, functions, and behaviours. The chief difference is that two advances mean that the process of evolution through selection acting on random mutations of the genetic material is now much more credible than it was at the beginning of the 20th century. First, we have a much richer body of data demonstrating the action of natural selection at every level of biological organization, from protein molecules to complex behaviour patterns. Second, we also now understand the mechanism of inheritance, which was a mystery to Darwin and Wallace. Many important aspects of heredity are now understood in detail, from how the genetic information is stored in the DNA, to how it controls the characteristics of the organism through the intermediacy of the proteins that it specifies and by regulating their levels of production. In addition, we now understand that many changes in DNA sequences have little or no effect on the functioning of the organism, so that evolutionary changes in sequences can occur by the random process of genetic drift. The technology of DNA sequencing enables us to study variation and evolution of the genetic material itself, and to use sequence differences to reconstruct the genealogical relationships between species.

This knowledge of heredity, and our understanding that natural selection drives the evolution of organisms' physical and behavioural characteristics, does not imply rigid genetic determination of all aspects of such characteristics. The genes lay down only the potential range of traits that an organism can exhibit; the traits which are actually expressed often depend on the particular environment in which an organism finds itself. In higher animals, learning plays a major role in behaviour, but the range of behaviour that can be learned is limited by the animal's brain structure, which is in turn limited by the animal's genetic make-up. This certainly applies across species: no dog will ever learn to talk (nor will humans be able to smell rabbits at a distance). Among humans, there is strong evidence for the involvement of both genetic and environmental factors in causing differences in mental characteristics; it would be astonishing if this were not the case in our species, as in other animals. Most variability among humans is between individuals within local populations, and differences between populations are far fewer. There is thus no basis for treating racial groups as homogeneous, distinct entities, much less for ascribing genetic 'superiority' to any one of them. This is an example of how science can provide knowledge to inform people's decisions on social and moral issues, although it cannot prescribe those decisions.

The characteristics which we regard as most human, such as our ability to talk and to think symbolically, as well as the feelings that guide our family and social relationships, must reflect a long process of natural selection that started tens of millions of years ago, when our ancestors started living in social groups. As we saw in Chapter 7, animals that live in social groups can evolve behaviour patterns that are not purely selfish, in the sense of promoting an individual's survival or reproductive success at the expense of another's. It is tempting to think that such characteristics as a sense of fairness towards others form part of our evolutionary heritage as a social animal, just as parental care of children surely represents evolved behaviour similar to that exhibited by many other animals.

We emphasize again that this does not mean that all details of people's behaviour are genetically controlled, or that they represent characteristics that increase human fitness. Moreover, there is great difficulty in conducting rigorous tests of evolutionary explanations for human behaviours.

Is there progress in evolution? The answer is a qualified 'yes'. More complex types of animals and plants have all evolved from less complex forms, and the history of life shows a general progression from the simplest type of prokaryote single-celled organism to birds and mammals. But there is nothing in the theory of evolution by natural selection to suggest that this is inevitable, and of course bacteria are still one of the most abundant and successful forms of life. This is analogous to the persistence of old, but still useful, tools such as hammers alongside computers in the modern world. In addition, there are many examples of evolutionary reduction of complexity, such as cave-dwelling species that have lost their sight, or parasites that lack the structures and functions needed for independent existence. As we have emphasized several times already, natural selection cannot foresee the future, and merely accumulates variants that are favourable under prevailing conditions. Increased complexity may often provide better functioning, as in the case of eyes, and will then be selected for. If the function is no longer relevant to fitness, it is not surprising that the structure concerned will degenerate.

Evolution is also pitiless. Selection acts to hone the hunting skills and weapons of predators, without regard to the feelings of their prey. It causes parasites to evolve ingenious devices to invade their hosts, even if this causes intense suffering. It causes the evolution of ageing. Natural selection can even cause a species to evolve such a low fertility that it becomes extinct when the environment takes a turn for the worse. Nevertheless, the vision of the history of life revealed in the fossil record, and in the incredible diversity of species alive today, gives a sense of wonder at the results of more than 3 billion years of evolution, despite the fact that this has all resulted 'from the war of nature, from famine and death', in Darwin's phrase. An understanding of evolution can teach us our true place in nature, as part of the immense array of living forms which the impersonal forces of evolution have produced. These evolutionary forces have given our own species the unique ability to reason, so that we can use our foresight to ameliorate the 'war of nature'. We should admire what evolution has produced, and take care not to destroy it through our greed and stupidity, but to preserve it for our descendants. If we fail to do this, our own species could become extinct, along with many other wonderful living creatures.

# Further reading

It is well worth reading *On the Origin of Species* by Charles Darwin (John Murray, 1859); the masterly synthesis of innumerable facts on natural history to support the theory of evolution by natural selection is dazzling, and much of what Darwin has to say is still highly relevant. There are many reprints of this available; Harvard University Press have a facsimile of the first (1859) edition, which we used for our quotations.

Jonathan Howard, *Darwin: A Very Short Introduction* (Oxford University Press, 2001) provides an excellent brief survey of Darwin's life and work.

For an excellent discussion of how natural selection can produce the evolution of complex adaptations, see *The Blind Watchmaker: Why The Evidence of Evolution Reveals a Universe without Design* by Richard Dawkins (W.W. Norton, 1996).

*The Selfish Gene* by Richard Dawkins (Oxford University Press, 1990) is a lively account of how modern ideas on natural selection account for a variety of features of living organisms, especially their behaviour.

*Nature's Robots. A History of Proteins* by Charles Tanford and Jacqueline Reynolds (Oxford University Press, 2001) is a lucid history of discoveries concerning the nature and functions of proteins, culminating in the deciphering of the genetic code.

Enrico Coen, *The Art of Genes. How Organisms Make Themselves* (Oxford University Press, 1999) provides an excellent account of how genes control development, with some discussion of evolution.

For an account of the application of evolutionary principles to the study of animal behaviour, see *Survival Strategies* by R. Gadagkar (Harvard University Press, 2001).

Richard Leakey and Roger Lewin, *Origins Reconsidered: In Search of What Makes Us Human* (Time Warner Books, 1993) gives an account of human evolution for the general reader.

J. Weiner, *The Beak of the Finch* (Knopf, 1995) is an excellent account of how work on Darwin's finches has illuminated evolutionary biology.

B. Hölldobler and E. O. Wilson, *Journey to the Ants. A Story of Scientific Exploration* (Harvard University Press, 1994) is a fascinating account of the natural history of ants, and the evolutionary principles guiding the evolution of their diverse forms of social organization.

For a discussion of the fossil evidence for the early evolution of life, and experiments and ideas on the origin of life, *Cradle of Life. The Discovery of Earth's Early Fossils* by J. William Schopf (Princeton University Press, 1999) is recommended.

*The Crucible of Creation* by Simon Conway Morris (Oxford University Press, 1998), which is beautifully illustrated, provides an account of the fossil evidence on the emergence of the major groups of animals.

## More advanced books (these assume an A-level knowledge of biology)

*Evolutionary Biology* by D. J. Futuyma (Sinauer Associates, 1998) is a detailed and authoritative undergraduate textbook on all aspects of evolution.

And a somewhat less detailed undergraduate textbook of evolutionary biology: *Evolution* by Mark Ridley (Blackwell Science, 1996).

*Evolutionary Genetics* by John Maynard Smith (Oxford University Press, 1998) is an unusually well-written text on how the principles of genetics can be used to understand evolution.

For a comprehensive account of the interpretation of animal behaviour in terms of natural selection, refer to *Behavioural Ecology* by J. R. Krebs and N. B. Davies (Blackwell Science, 1993).

# “牛津通识读本”已出书目

古典哲学的趣味
人生的意义
文学理论入门
大众经济学
历史之源
设计，无处不在
生活中的心理学
政治的历史与边界
哲学的思与惑
资本主义
美国总统制
海德格尔
我们时代的伦理学
卡夫卡是谁
考古学的过去与未来
天文学简史
社会学的意识
康德
尼采
亚里士多德的世界
西方艺术新论
全球化面面观
简明逻辑学
法哲学：价值与事实
政治哲学与幸福根基
选择理论
后殖民主义与世界格局
福柯
缤纷的语言学
达达和超现实主义
佛学概论
维特根斯坦与哲学
科学哲学
印度哲学祛魅
克尔凯郭尔
科学革命
广告
数学
叔本华
笛卡尔
基督教神学
犹太人与犹太教
现代日本
罗兰·巴特
马基雅维里
全球经济史
进化
性存在
量子理论
牛顿新传
国际移民
哈贝马斯
医学伦理
黑格尔
地球
记忆
法律
中国文学
托克维尔
休谟
分子
法国大革命
丝绸之路
民族主义
科幻作品
罗素
美国政党与选举
美国最高法院
纪录片
大萧条与罗斯福新政
领导力
无神论
罗马共和国
美国国会
民主
英格兰文学
现代主义
网络
自闭症
德里达
浪漫主义

批判理论
电影
俄罗斯文学
古典文学
大数据
洛克
幸福
德国文学
戏剧
腐败
医事法
癌症
植物
儿童心理学
时装
现代拉丁美洲文学
卢梭
隐私
电影音乐